Dark Energy
and
Human Consciousness

Also by Tom Cahalan
Beyond the Horizon: Where Science Finds God
ISBN 978-0-9574289-0-4, Published in 2012.

In this book I describe the creation and structure of the universe. I look at biological evolution and the development of human consciousness. I explain the interconnectedness of all aspects of reality, from the source of the universe to human consciousness. I discuss the influences that religion, politics, and science have on human consciousness and, consequently, on the world in which we live.

Finally, I suggest that the scientific mystery of black holes, dark energy, and dark matter and the religious mystery of God the Father, God the Son, and God the Holy Spirit are synonymous.

This is a book for everyone who has ever wondered about the mystery that is God.

Dark Energy
and
Human Consciousness

Humanity's Path to Freedom

Tom Cahalan

BALBOA.
PRESS
A DIVISION OF HAY HOUSE

Balboa Press books may be ordered through booksellers or by contacting:

Balboa Press
A Division of Hay House
1663 Liberty Drive
Bloomington, IN 47403
www.balboapress.com
1 (877) 407-4847

Because of the dynamic nature of the Internet, any web addresses or
links contained in this book may have changed since publication and
may no longer be valid. The views expressed in this work are solely those
of the author and do not necessarily reflect the views of the publisher,
and the publisher hereby disclaims any responsibility for them.

The author of this book does not dispense medical advice or prescribe the use
of any technique as a form of treatment for physical, emotional, or medical
problems without the advice of a physician, either directly or indirectly. The
intent of the author is only to offer information of a general nature to help
you in your quest for emotional and spiritual well-being. In the event you use
any of the information in this book for yourself, which is your constitutional
right, the author and the publisher assume no responsibility for your actions.

Any people depicted in stock imagery provided by Thinkstock are models,
and such images are being used for illustrative purposes only.
Certain stock imagery © Thinkstock.

Printed in the United States of America.

ISBN: 978-1-4525-1594-6 (sc)
ISBN: 978-1-4525-1596-0 (hc)
ISBN: 978-1-4525-1595-3 (e)

Library of Congress Control Number: 2014909681

Balboa Press rev. date: 07/18/2014

To my wife Liz; my children Orlaith, Eibhlis, Tomás, Tadhg, and Micheál; and my grandson Jack

Tom Cahalan qualified as a mechanical engineer at the College of Technology, Bolton St, Dublin, in 1974. He worked in Canada and the USA for eight years afterward and returned to Ireland in the early 1980s. He lectured in the Dublin Institute of Technology until the mid 1980s. Afterward, with two colleagues, he founded an engineering company that specialised in the design, start-up, and commissioning of production facilities for the electronics and pharmaceutical industries. They sold the company in 2000.

For the past twelve years he has been researching the interconnectedness among the universe, life, mind, and human consciousness.

He published *Beyond the Horizon: Where Science Finds God* in 2012. This book explains the interconnectedness of all aspects of reality, from the source of the universe to human consciousness. It also discusses the influences that religion, politics, and science have on human consciousness and, consequently, on the world in which we live.

Tom lives in Clane, Co Kildare with his wife and family.

Contents

Preface

In our world today, 870 million people are imprisoned in a life of poverty and starvation. Of those 870 million people, about 21,000 people die every day of hunger or hunger-related causes. I say they are imprisoned, because these people have no way of escaping from the prison of their lives. So who are the prison guards?

The prison guards are the 0.111 per cent of humanity who control 81 per cent of the world's wealth. This inequitable distribution of wealth is largely responsible for the high level of poverty and starvation in our world. However, blaming the rich for being rich or the poor for being poor is a futile exercise. The only hope of escaping from this terrible reality is to understand why it exists.

The main reason for the existence of this inequitable distribution of wealth is the culture of selfishness, corruption, abuse of power, and greed, which dominates the political, religious, public service, and financial institutions worldwide. For ease of reference, I will give this culture of selfishness, corruption, abuse of power and greed the acronym SCAG. It is futile to blame the leaders of these institutions for the dominance of SCAG because few of us can say that we would act any differently if we occupied these positions of power. Our only hope is to understand why this SCAG culture is so dominant in the institutions where power and control come with the job.

In this book I suggest that the fundamental reason for this SCAG culture is the innate quality of selfishness that is in every human being. When selfishness gains a foothold in our consciousness, greed quickly follows, and once we enter that terrain, then corruption and abuse of power almost inevitably results whenever the opportunity presents itself. Of course, the opportunity usually presents itself to people in power, which is why those of us who are not in power should not be too self-righteous or condemn these people too quickly. Why is selfishness an innate quality in every human being, and why can't we move past it?

It is a scientific fact that the natural propensity of matter is to evolve to states of lower complexity. Since humans have a physical aspect, they also have a force pulling them to lower complexity. If we consider human consciousness in terms of complexity, the lowest level of complexity is to offer care and compassion only to oneself. Examples of increasing levels of complexity are to offer care and compassion to family, community, society, country, and eventually to all of humanity; in other words, lower complexity equals lower inclusion, and higher complexity equals higher inclusion.

The innate urge of the physical aspect to move to lower complexity means that we naturally offer care and compassion only to ourselves. Therefore, this physical aspect of our being is naturally pulling us toward selfishness – to taking care of ourselves first and last.

If life is purely physical, then we have no escape from selfishness, so SCAG will always dominate in our world. However, in this book I present a strong argument *based on scientific concepts* that suggests that there is a non-physical aspect to all life, including human life. I also argue that this non-physical aspect has a natural propensity to move to higher complexity or higher inclusion. Therefore if we acknowledge and nurture this non-physical aspect of ourselves, it will balance the natural tendency of the physical and help humanity to evolve to a more enlightened state where care and compassion are extended *to all*, and just as importantly, *by all*.

In this enlightened state, selfishness, corruption, abuse of power, and greed will no longer dominate, the inequitable distribution of wealth will no longer be a reality, and poverty, starvation, conflict, and war will cease to exist.

This book uses some novel concepts that may challenge the belief systems of many readers. However, I ask you to read them with an open mind and let them rest awhile in your consciousness. I believe that, as they sink in, you will realise that they make a lot of sense.

While this book uses some complicated scientific concepts, it is written so all readers can understand. This is important because humanity cannot move to a more enlightened world view until a critical mass of people join in the effort.

If we succeed in the pursuit of the goals suggested in this book, our reward will be to live in a world without SCAG and, consequently, without wars, social discrimination, poverty, or hunger. That possibility should be enough to convince everybody that this book is a must-read for all.

It will also be of interest to a number of specific groups:

- Physicists will find it enlightening, as it logically puts in context many scientific mysteries of the present time, such as black holes, dark energy, dark matter, particle entanglement, and wave-particle duality.
- Geneticists and biologists will benefit from the new concepts suggested regarding the origin and evolution of life.
- Philosophers and theologians will be presented with new concepts that show the interconnectedness of all aspects of reality, from the source of the universe to human consciousness.

- The leaders of religious institutions will benefit from the suggestion that their institutions are the source of much suffering and bloodshed.

In summary, this book has the potential to change forever the way we live our lives.

CHAPTER 1

The Present Reality

Preamble

When we look at our world, should we be pleased with what we see, or should we be disappointed or even outraged with the world we have created?

There is no doubt that the standard of living, particularly in the western world, has substantially improved in the last three centuries. This improved standard of living has been driven primarily by science and technology, starting with the Industrial Revolution in the mid-eighteenth century and moving on to electrification in the mid-nineteenth century, the internal combustion engine in the late nineteenth century, air travel in the early twentieth century, space travel in the mid-twentieth century, and the IT revolution in the late twentieth century, to name but a few moves forward. However, in spite of all this progress, has the quality of life for the general public improved over this time?

Quality of life should not be confused with standard of living, which is based primarily on income. The term 'quality of life' refers to the general well-being of individuals and societies; it relates not just to wealth and employment but also to the physical infrastructure, physical and mental health, education, recreation, leisure time, social belonging, freedom, human rights, personal security, social justice, equality for all, and so on. In general, a good quality of life is a prerequisite for the achievement of happiness. Happiness is subjective and difficult to measure, but it is certain that it does not necessarily increase in proportion to increasing income. As a result, one's standard of living should be taken as only one of the many components that influence happiness.

However, a certain level of wealth is necessary in order to achieve a good quality of life. To quote a line from a speech former US President Bill Clinton made during one of his visits to Ireland, 'Money is not everything, but it's up there with oxygen.' Mr. Clinton's one-liner rightly suggests that we need money to survive. However, the similarities between money and oxygen end there. In the case of oxygen, there is an adequate supply available to all people, and nobody can hoard it or build up a supply in case he or she might need it in the future. Sadly, that is exactly what is happening with money; the few with the power control the money supply, leaving the vast majority with an inadequate amount on which to survive.

Distribution of Wealth

The distribution of wealth is a comparison of the wealth of various members or groups in a society or country and should not be confused with the distribution of income. Wealth consists of those items of economic value that an individual owns, while income is an inflow of items of economic value. A detailed analysis of these topics would fill many books and would only confuse the central points I wish to convey in this book, so my intention here is merely to explain that the distribution of wealth in the world is not very equitable.

In the US, 5 per cent of the population owns almost 62 per cent of the wealth, while the bottom 40 per cent owns only 0.2 per cent – not a very equitable distribution. This problem is even more lopsided if we look at the global situation, where 0.111 per cent of the population owns 81 per cent of the wealth, while the bottom 99.889 per cent owns only 19 per cent.

This inequitable distribution of wealth creates a global society where a very small percentage of people live in luxury; a large working class struggles to pay its mortgages and taxes, educate its children, and make ends meet; and a large lower class lives in poverty. Of course, there are many sub-levels in each of these three categories, but it is an accurate summation of the overall situation.

Quality of life

In addition to the inequitable distribution of wealth – and sometimes because of it – the quality of life for the vast majority of humanity is much lower than that you would expect in a civilized world. Some examples of this unacceptable quality of life follow.

Hunger

The United Nations Food and Agriculture Organization estimates that nearly 870 million people, or one in eight of the 7.1 billion people in the world, suffered from chronic undernourishment in 2010–2012. Of these 870 million undernourished people, 852 million lived in developing countries, and approximately 18 million lived in developed countries.

According to the United Nations, about 21,000 people die every day of hunger or hunger-related causes – almost fifteen people every minute, ten of whom are children. Children are the most vulnerable victims of under-nutrition. Poorly nourished children suffer up to 160 days of illness each year. Poor nutrition is the main cause of 5 million child deaths each year, which is almost ten deaths every minute.

Poverty is the principal cause of hunger. Hunger can lead to even greater poverty by reducing people's ability to work and learn, thus leading to even greater hunger, a vicious cycle.

Poverty

Poverty is a major problem in the world today. There are fundamentally two types of poverty: (1) absolute poverty (or destitution), where people lack the basic human needs, such as water, food, healthcare, clothing, shelter, and education; and (2) relative poverty, which refers to economic inequality in a particular country or society and is the most useful measure for ascertaining poverty rates in wealthy developed nations.

The threshold for absolute poverty is difficult to establish. For example, the absolute poverty line in the US was $15.15 per day in 2010 (for a family of four), while in India it was $1.00 per day. The World Bank has set a threshold of $1.25 per day, although some experts disagree with this number. World Bank data (2011) reveals that, in 2008, 22.4 per cent of the world's population lived on less than €1.25 per day

Relative poverty is the most useful indication of poverty rates in wealthy developed nations. Relative poverty is usually measured as the percentage of population with income less than some fixed proportion of the median income in a particular country or society. It is a reflection of the cost of social inclusion and equality of opportunity in a specific location or society.

The principal cause of poverty (and, consequently, of hunger) is how the economic and political systems in the world operate. Control over food, resources, wealth, and income is normally based on the military, political, and economic power that typically ends up in the hands of a minority who live in luxury, while those at the bottom barely survive. In fact, many do not survive.

War and Conflict

War and conflict have been the cause of immense suffering, bloodshed, poverty, hunger, and deprivation for thousands of years. For example, the approximate number of fatalities in World War I was 16 million; World War II, 60 million; the Korean War (1950–1953), at least 1.2 million; the Vietnam War (1955–1975), at least 2 million; the Colombian conflict (1964–present), 600,000; the conflict in Afghanistan (1978–present), 2 million; the Somali civil war (1988–1991), 500,000; the Israeli-Palestinian conflict (1948), 20,000; the Iraq War (2003–2013), 500,000; and the war in Darfur (2003–2010), up to 450,000.

When you add to these figures the numbers of wounded, the societies destroyed and left destitute, the millions of people left homeless, and the millions of children left without parents, the awful savagery of these conflicts becomes even more apparent. Many conflicts, such as those in Egypt, Syria, and Iraq, continue unabated.

When the extent of hunger and poverty unrelated to war are added to the consequences of the wars and conflicts, it is clear that the quality of life for the vast majority of the world's population is far below that which you would expect from a civilized human race. In fact if you could look down on the Earth and observe all that is happening, you would conclude that the human race is an extremely dysfunctional family.

As you would expect, there are two opposing views regarding poverty and deprivation: one from the 'haves' and one from the 'have-nots'. The haves are usually right of centre politically, who can be loosely described as 'conservative'. On the other side are those who are generally left of centre politically, who can be loosely described as 'liberal'.

Many conservatives believe that the poor are poor because they are lazy and unwilling to take personal responsibility. They normally recommend that we return to traditional 'family values', that individuals assume more personal responsibility, that we tighten up slack moral standards, encourage a strong work ethic and reward achievement, and so on.

On the other hand, many liberals believe that the poor are poor because the system has let them down or that they are oppressed by society. Therefore, the liberal solution to the problem is based on social change, such as a redistribution of wealth, changes to social institutions, taxing the rich to help the poor, and so on.

Of course, there is some validity in both arguments. However, it is necessary to understand the fundamental cause of the problem before we can propose meaningful solutions. The remainder of this chapter gets to the core of the problem and outlines its root cause.

The Present Paradigm

A paradigm is a set of assumptions, concepts, values, and practices that constitute a way of viewing reality. While each society may have its own particular paradigm, they all have SCAG in common because selfishness is a human trait. When selfishness dominates in a person or society, it is quickly followed by greed, and once we enter that terrain, corruption and abuse of power naturally follow when the opportunity presents itself.

Selfishness is a natural trait in every individual, but not all people are corrupt or abuse their power; there are many honest, hardworking individuals in every walk of life. However, it is fair to say that a *critical mass*[1] of people in all political, religious, financial, and public service institutions worldwide are dominated by SCAG, and this is the fundamental reason for the inequitable distribution of wealth, the high incidences of poverty and starvation, and the many wars. The best way to understand this statement is to imagine what life would be like if everybody lived his or her life with an attitude of love, care, and compassion for fellow humans. Imagine what life in your family and local community would be like if this were the case. Then imagine if the political, religious and public service institutions in your country operated based on an attitude of love, care, and compassion. Finally, imagine what life would be like if this attitude spread to every political, religious, and public service institution in the world. I think you would agree that, if this were the case, we would be living in a much happier world – a world with no wars, no poverty, no starvation, and no social, racial, or religious discrimination.

However, back in the real world, SCAG in our political, public service, financial, and religious institutions has a debilitating effect on every society on our planet:

---

[1] Critical mass *is a sociodynamic term to describe the existence of sufficient momentum in a social system such that the momentum becomes self-sustaining and fuels further growth.*

In third-world and emerging countries it is the root cause of widespread poverty, starvation, conflict, and war.

In developed countries, it is the cause of significant poverty and starvation and also approximately doubles the tax burden on the hard-pressed working class.

This last point concern all tax payers. Cumbersome regulations, procedures, and guidelines are necessary in order to contain SCAG, not just in the government and the public service, but also at every level of society. While these regulations, procedures, and guidelines are essential just to contain SCAG at an 'acceptable' level, they drastically increase the cost and reduce the efficiency of providing all public services. Additionally, as more instances of corruption are uncovered, more layers of regulations, procedures, and bureaucracy are added and the cost of containing SCAG increases still further. Unfortunately the tax payer pays the price.

SCAG has existed for thousands of years – probably ever since humans started to think and reason. It exists at all levels of society, not just in the political, religious, financial, and public service institutions. Evidence can be seen in fraudulent behaviour by highly paid managers and executives in many companies and corporations, fraudulent insurance claims by people at all levels, fraudulent social welfare claims, poor work ethic at all levels of society, the prevalence of all types of criminal behaviour, the many people who are focussed solely on personal gain and take little or no time to help out at community level, and so on. Because SCAG dominates our political, religious, financial, and public service institutions, it has serious consequences for humanity.

The primary source of the SCAG phenomenon is the innate trait of selfishness that exists in every human being. Selfishness is a natural trait in all human beings. Understanding SCAG requires that we understand why this is so, why selfishness dominates to such an extent, whether there is another innate trait in human beings that can balance it, and how we can harness this 'balancing trait' to help lead us to a more enlightened paradigm. In other words we require a better understanding of life and how it originated and evolved.

However, before we can understand life, we must first understand how the universe originated and evolved. Therefore the universe is discussed in the next chapter and life is discussed in the following chapter.

CHAPTER 2

The Universe

The universe is everything that exists physically, the entirety of space and time, all forms of matter, energy, and momentum, and the physical laws and constants that govern them. The age of the universe is believed to be approximately 13.7 billion years, and scientists estimate that the diameter of the *observable* universe is at least 93 billion light years.

I use the term 'scientist' to include geneticists, biologists, physicists, mathematicians, astronomers, and practitioners in all professions involved in the logical research of life, matter, and energy who study the physical universe.

The Earth is part of the solar system. The solar system exists in the Milky Way galaxy. The sun is one of the stars in the Milky Way, and the only reason it looks bigger than all the other stars, is that it is so close to us. There are 100 billion (10^{11}) stars in the Milky Way and 100 billion galaxies in the observable universe, so there are approximately 10^{22} stars in the observable universe, and many of them have planetary systems similar to our solar system. Therefore, the possibility that life exists out there somewhere must be very high.

The Composition of the Universe

Scientists estimate that the universe is made up of 4 per cent ordinary matter, 26 per cent dark matter, and 70 per cent dark energy. (These figures may vary somewhat depending on the source.) My own belief is that the percentage of ordinary matter is much less than 4 per cent of the total and that it may actually be infinitesimal. However, for the purposes of presenting my arguments, I use the 70/26/4 per cent estimates.

The language of physics is mathematics, but physicists are unable to describe the nature of dark energy or dark matter mathematically, to observe them, or to examine them experimentally. Therefore scientists have no knowledge regarding the nature of dark energy or dark matter. How do they know that dark energy and dark matter exist?

Dark Matter

Galaxy clusters are collections of galaxies that are bound together by gravity. However, because of their high velocities, their mutual gravitational attraction (calculated on the basis of 'ordinary' matter) seems insufficient for them to remain gravitationally bound, so there must be an invisible mass component. The invisible mass component, sometimes called the 'missing mass component', whose nature is still unknown, is given the name *dark matter.*

Dark Energy

Dark Energy is a hypothetical form of energy that permeates all of space. It is the most popular way of explaining recent observations that the universe appears to be expanding at an accelerating rate.

Therefore, all the scientific knowledge we possess today refers to only 4 per cent of reality, which is a sobering thought. Scientists have been aware of the existence of dark matter since 1932 and dark energy since 1998, yet they are unable to describe them mathematically.

The Origin of the Universe – The Scientific View

Most scientists believe that the universe started with the 'Big Bang', the primeval explosion that brought all space and time, all matter and energy, into being. The Big Bang theory postulates that the universe was only a few millimetres in diameter about 13.7 billion years ago. However, it is not possible for scientific theory to extrapolate what might have existed close to time zero, or prior to the Big Bang, since all the theories of cosmology break down prior to time zero.

The Evolution of the Universe

Scientists divide the evolution of the universe into two eras: the radiation era and the matter era. They break the radiation era down further into five epochs and the matter era into three epochs (Table 2.1).

Era	Epoch	Time (after Big Bang)	Main Events
Radiation Era	Planck	0 to 10^{-43} seconds	Unknown physics. Four fundamental forces were united.
	Grand Unification	10^{-43} to 10^{-35} seconds	Gravity decoupled from the still unified strong, weak and electro-magnetic forces. Quarks and anti-quarks form.
	Hadron	10^{-35} to 10^{-4} seconds	The remaining forces separated. Hadrons formed and dominated the mass of the universe.
	Lepton	10^{-4} to 10^{2} seconds	Leptons formed and dominated the mass of the universe
	Nuclear	10^{2} s to 10^{3} years	Nuclei of atoms formed and helium was created. The radiation era ended at year 1,000 and the matter era commenced.
Matter Era	Atomic	10^{3} to 10^{6} years	Atoms form. Matter begins to dominate.
	Galactic	10^{6} to 10^{9} years	Galactic and large-scale structures form
	Stellar	10^{9} years to present	All galaxies have formed. Stars and planets continue to form.

Table 2.1: Evolutionary Eras and Epochs of the Universe

Planck Epoch

The physics of the Planck epoch are unknown, but scientists believe that all four fundamental forces – strong nuclear, electromagnetic, weak nuclear, and gravitational force – were united at that time. This essentially means that scientists have no knowledge regarding the 'stuff' that existed in the Planck epoch.

Grand Unification Epoch

In the Grand Unification epoch quarks and anti-quarks emerged from the 'stuff' of the Planck epoch, and gravity emerged as a fundamental force. Quarks are believed to be fundamental particles – that is, particles that are not formed by a composite of smaller particles.

Hadron Epoch

During the Hadron epoch the remaining three fundamental forces emerged. Shortly after 10^{-35} seconds the strong force decoupled from the still unified electro-weak force. At approximately 10^{-12} seconds the weak and electromagnetic forces separated.

Hadrons are formed from quarks held together by the strong force. There are two types of hadrons; baryons (e.g., protons and neutrons, which are made of three quarks) and mesons (e.g., pions, which are made of one quark and one anti-quark). By the end of the Hadron epoch, free quarks had disappeared. Most of the hadrons and anti-hadrons were then eliminated in annihilation reactions, leaving a small residue of hadrons.

Lepton Epoch

A lepton is an elementary particle that does not interact with the strong force. The best known of all leptons is the electron, which governs nearly all of chemistry, as it is found in atoms and is directly tied to all chemical properties. There are two main classes of leptons: charged leptons (also known as electron-like leptons) and neutral leptons (better known as neutrinos). Approximately ten seconds after the Big Bang, the temperature of the universe fell to the point at which lepton/anti-lepton pairs were no longer created, and they began to annihilate each other, leaving only a small residue of leptons.

The evolution of the universe then continued with the formation of nuclei of atoms, atoms, molecules, matter, galaxies, stars, planets, and moons.

Scientists have a reasonable knowledge of the sequence of events after the formation of quarks occurred during the Grand Unification epoch, but they have little knowledge of events between time zero and 10^{-43} seconds after the Big Bang. Another early event that scientists do not fully understand is the inflationary period. It is unknown exactly when the inflationary period actually started, but it is thought to have ended approximately 10^{-32} seconds after the Big Bang. According to Alan Guth, a scientist at the Massachusetts Institute of Technology, during this time the radius of the universe increased by 10^{30} times in only a fraction of a second.

The four forces, known as forces of nature, had a vital impact on how the universe evolved and are known as the strong nuclear, electromagnetic, weak nuclear and gravity. A short description of each follows:

The Four Forces

The strong nuclear is the strongest of the four forces of nature. It holds together the protons and neutrons, which are made of still tinier particles called quarks. This force also binds the protons and neutrons inside the nucleus of an atom.

The electromagnetic is a long-range force and the second-strongest of the four forces of nature. The electromagnetic force holds atoms and molecules together. It acts only on particles with electric charge; it is repulsive between charges of the same sign, and attractive between charges of opposite sign. This means the electric forces between large bodies cancel each other out, but on the scale of atoms and molecules, they dominate. Electromagnetic forces are responsible for all of chemistry and biology.

The weak nuclear force plays a vital role in the formation of the elements in stars and the early universe. It is responsible for the radioactive decay of subatomic particles. It also initiates the process known as hydrogen fusion in stars.

Gravity is the weakest of the four forces of nature, yet it is the dominant force in the universe for shaping the large-scale structure of galaxies, stars, and so on. It is a long-range force that acts on everything in the universe, so for large bodies the gravitational forces all add up and can dominate all other forces. Gravitational force is the means by which objects that have mass attract each other.

The relative strengths of these forces are summarised in Table 2.2.

Force	Relative Strength	Function	Comment
Strong Nuclear	10^{38}	Binds Nucleus	Short Range (10^{-15} m)
Electromagnetic	10^{36}	Binds Atoms and molecules	Long Range (∞)
Weak Nuclear	10^{25}	Causes Radioactivity	Short Range (10^{-18} m)
Gravitational	1	Binds Earth to Sun, Binds Galaxies	Long Range (∞)

Table 2.2: Summary of the Four Forces of Nature

The strengths of the four forces are central to the evolution of the universe. For example scientists have calculated that minute changes in the strength of the strong force or the electromagnetic force, would destroy either all the carbon or all the oxygen in every star, and life as we know it could not exist.

In 1931, Monseigneur Georges Lemaître proposed in his *hypothèse de l'atome primitif* (hypothesis of the primeval atom) that the universe began with the 'explosion' of the 'primeval atom'. Around the same time James Jeans proposed an alternative model called the Steady State theory. This theory was revised in 1948 by Fred Hoyle and others. It was Hoyle who coined the name of Lemaître's theory, referring to it as 'this big bang idea' in March 1949. Most scientists now believe that all matter, energy and space in the universe was once squeezed into an infinitesimally small volume, which erupted in a cataclysmic 'explosion' that has become known as the Big Bang. The Big Bang attempts to explain how the universe developed from a tiny, dense state into what it is today, but this theory is merely a model to convey what happened in the beginning, not a description of an actual explosion. An explosion is slightly misleading in that it conveys the image that the Big Bang was triggered in some way and at some particular point in space and time. However, the same pattern of expansion would be observed from anywhere in the universe, so there is no particular location in our present universe which could claim to be the origin. The Big Bang really describes a very rapid expansion or stretching of space itself, rather than an explosion in pre-existing space

An easy way to visualise the Big Bang is to imagine that everything that exists in the universe today was once squeezed into a sphere only a few millimetres in diameter. As this small sphere rapidly expanded, all the components in that sphere moved farther apart, just as is happening today with the expansion of the universe.

However, for two primary reasons, this model regarding the origin of the universe makes little sense.

Reason 1

The Big Bang model predates scientists' realisation that dark matter and dark energy exist, so it must be based on a purely physical universe, which is only 4 per cent of the whole. To suggest that the origin of the universe is based on 4 per cent of reality makes little sense.

Consider the analogy of the water cycle. In the water cycle, the sea is the source. Water vapour evaporates from the sea and fills the atmosphere. When conditions are right, clouds form from the water vapour. The clouds then either return to the sea as rain or evaporate back into the atmosphere again. For the purposes of this analogy, imagine the clouds are the physical (manifest) universe, the water vapour is the non-physical (unmanifest) essence and the sea is the source. In my view, saying that matter is the source of the universe is akin to saying that clouds are the source of the sea.

Reason 2

Scientists believe that all four forces were united in the Planck epoch. If the four forces were united at the present time, there would be no gravity to keep all the celestial bodies in existence, there would be no electromagnetic force to bind atoms and molecules, and there would be no strong force to bind the elementary particles or to bind the nuclei of atoms. Therefore, all matter would return to elementary particles, such as electrons and quarks. The elementary particles would then return to the 'stuff' that existed in the Planck epoch, which would then account for 100 per cent of the universe.

However, there is no logical argument to suggest that the 'stuff' of the Planck epoch would then return to an infinitesimally small sphere. In my view the concept of this 'infinitesimally small sphere' is rooted in the belief that the universe is a purely physical phenomenon. However, since scientists now believe that only 4 per cent of the universe is physical, I believe that we should look to the 96 per cent, rather than the 4 per cent, to understand the origin of the universe.

The Origin of the Universe –
An Alternative Model

The aspect that dark energy, dark matter, and the state that existed during the Planck epoch (Table 2.1) have in common is that their nature is not understood by science. This leaves open the real possibility that dark energy and dark matter existed in the Planck epoch and may be the source of elementary particles like quarks – and consequently the source of all matter. This view makes much more sense than the scientific view since dark energy and dark matter constitute 96 per cent of the 'stuff' that makes up the universe.

However, dark energy and dark matter are two separate phenomena, and there can only be one source.

Black Holes

A black hole is a region of space in which the gravitational field is so powerful that nothing, not even light, can escape its pull. Scientists believe that it is the result of the deformation of space-time caused by a massive and very compact mass. It is called 'black' because it absorbs all the light that comes toward it, reflecting nothing, just like a perfect black body in thermodynamics. Around a black hole there is an undetectable surface that marks 'the point of no return', called an event horizon.

The No-Hair Theorem postulates that a black hole can be characterized completely by only three externally observable parameters: mass, angular momentum, and charge. All other information about the matter that formed a black hole disappears behind the black hole event horizon and is permanently inaccessible to external observers.

The black hole information paradox suggests that physical information could disappear in a black hole, allowing many physical states to evolve into exactly the same state. However, this view violates a commonly assumed tenet of science that states that, in principle, complete information about a physical system at one point in time should determine its state at any other point in time. Could the explanation of this black hole information paradox (allowing many physical states to evolve into exactly the same state) lie in the fact that complete unity exists inside a black hole? In other words, everything that is drawn into a black hole is transformed by its enormous internal forces into a 'homogeneous essence' or a complete 'unity'. Should this be the case, there is a strong logical argument to suggest that black holes are the source of dark energy and dark matter. How do scientists know that black holes exist?

Einstein's theory of general relativity implies the existence of black holes as an end state for massive stars. Despite its invisible interior, a black hole can be observed through its interaction with other matter; for example, the presence of a black hole can be inferred by tracking the movement of a group of stars that orbit a region in space that appears to be empty.

Alternatively, when gas falls into a stellar black hole from a companion star, the gas spirals inward, heating to high temperatures as it approaches the event horizon and emitting large amounts of radiation that can be detected by earthbound and earth-orbiting telescopes. Astronomers have identified numerous stellar black holes and have also found evidence of super-massive black holes at the centre of many galaxies. After observing the motion of nearby stars for many years, astronomers have found convincing evidence that a super-massive black hole of more than 4 million solar masses is located in the centre of our own Milky Way galaxy.

The argument for an alternative view of the origin of the universe begins with surmising what might have existed before time zero (before the Big Bang). If a physical universe existed prior to the Big Bang, all of the ordinary matter (the 4 per cent) is likely to have returned to dark matter, triggered by the unification of the four forces (as scientists believe was the case during the Planck epoch). This unification would allow all matter to return to its source, which I contend is dark matter. Dark matter and dark energy may then have returned to a black hole. Therefore, all that would have existed at time zero was one massive black hole.

Based on this reasoning, there is a much more logical model for the origin of the universe:

- Just prior to time zero, the complete manifest universe (the 4 per cent)—if there was a one prior to time zero – plus all dark energy and dark matter had returned to one massive black hole.
- At time zero dark energy and dark matter emanated from the black hole.
- Quarks became manifest from dark matter, and manifestation of the physical universe continued, in accordance with scientific theory, with the formation of hadrons, leptons, nuclei of atoms, atoms, molecules, matter, galaxies, stars, planets, and moons – all the way up to the universe as we know it today.
- Dark energy proceeded with infinite speed to fill all of space, which may well explain the inflationary period that ended about 10^{-32} seconds after the Big Bang and during which the radius of the universe is believed to have increased by 10^{30} times in a fraction of a second.

I suggest that the universe is a constant cycle of emanation, manifestation and return. The complete physical universe (the 4 per cent) may sometime return to dark matter, which may return to a black hole with dark energy, and the cycle will repeat. The strength of this concept is that it involves 100 per cent of that which constitutes the universe in its origin and evolution, whereas only 4 per cent is included in the current scientific view.

The concept that quarks and other elementary particles emerged from dark matter makes it easier to understand many scientific mysteries, such as particle entanglement. I refer to entanglement as a mystery because scientists know that it is real but do not yet understand why or how it occurs.

Imagine there are two entangled particles, A and B, so the characteristics of particle A are opposite to those of B. For example, if A has a clockwise spin, then B has a counter-clockwise spin. Now imagine that particle B is removed to a location thousands of miles from A and that A is then entangled with a third particle, C, which has the effect of reversing A's spin to counter-clockwise. Scientists have demonstrated that, at the instant C is entangled with A, B (in the remote location) immediately assumes the opposite characteristics of A. Since there is no obvious link between A and B, scientists do not understand this phenomenon.

However, if we assume that these elementary particles (A, B, and C) emerged from dark matter, then it is reasonable to assume that they were encoded by dark matter. In addition, since dark matter permeates all of space, all three particles are immersed in dark matter; therefore, dark matter is the link between particles A and

B. Furthermore, since velocities in dark matter can be infinite, it's reasonable to assume that the characteristics of B changed at the same instant that the characteristics of A changed.

The concept that elementary particles emerge from dark matter also makes it possible to explain the counter-intuitive results of the double-slit experiment, which demonstrates that photons can behave as both particles and waves.

I believe that 100 per cent of that which constitutes the universe influences not only the origin and evolution of the universe itself, but also the origin and evolution of the life in it. I present an argument in support of this belief in the next chapter. First, though, I want to discuss why science is finding it so difficult to understand dark energy and dark matter. Science understands particles, waves, momentum, forces and their characteristics – that is, the physical, manifest, universe, the 4 per cent. However, dark energy and dark matter are prior to matter – prior to the physical, so they are non-physical or non-manifest. I use the words 'physical' and 'non-physical' in this context in the remainder of this book.

The non-physical (non-manifest) state does not lend itself to the science of the physical universe – that is, to the science of the 4 per cent. Mathematics is the language of physics, and it must have something (some 'thing') to describe before it can be applied. However, in the non-physical state, nothing (no 'thing') exists, so mathematics are not applicable.

It is for this reason that I believe scientists will not be able to describe black holes, dark energy, or dark matter with present-day mathematics. They will first have to find a 'new mathematics' that can be applied to the non-physical state.

Chapter 3

The Origin and Evolution of Life

Nobody knows how life was first breathed into the first atom, molecule, or cell; all biologists can do is to suggest the most reasonable possibilities by analysing the oldest DNA available. Scientists postulate that life emerged from non-life via natural processes about 3.7 billion years ago. It then evolved (and continues to evolve) to higher levels of complexity based on mutation, natural selection, and time. Science also suggests that all mutations are the result of chance, which suggests that scientists regard life as a purely physical phenomenon. In other words, they regard the physical universe (the 4 per cent) as a closed system, and that dark energy and dark matter have no influence on the origin or evolution of life.

Although this belief is not logical, it is crucial to our understanding of SCAG. Since all the major advances that are responsible for the wealth that exists have been driven by science and technology, the minority of people who control all the wealth in our world will listen only to the scientific view, if they listen to anybody at all. However, based on the scientific view of life, selfishness and SCAG are an inevitable reality.

If life is purely physical, then the natural propensity of matter is crucial to the evolution of life. Matter has a natural propensity to evolve toward lower complexity, as the second law of thermodynamics and the concept of entropy explain. The second law of thermodynamics is an expression of the universal principle of decay that is observable in nature and measured and expressed in terms of a physical property called entropy. The second law states that the entropy of an isolated system can never decrease; if it is not at thermodynamic equilibrium, it will tend to increase over time, reaching a maximum value when equilibrium is reached. Therefore, the material world, left to its own devices, will always move toward increased disorder, increased entropy, and lower complexity.

A simple example of this principle is a steel pipe, which started its life as iron ore in the Earth. Through processing and the expenditure of a large amount of energy, the iron ore is transformed into steel pipes, which are at a higher level of complexity than iron ore. However, the natural tendency of the pipe is to return to its original state of iron ore, which is the force behind corrosion. Therefore, if a steel pipe is left unprotected for a sufficiently long period, it will return to the iron ore state again, a state of lower complexity.

It also makes sense that the natural propensity of matter is toward lower levels of complexity once we accept that everything in nature has an innate urge to return to its source. Matter emerged from dark matter – from lower complexity (e.g., quarks, hadrons, leptons) to higher complexity (e.g., galaxies, stars, planets). Therefore, if the emergence was from lower complexity to higher complexity, then the return must be from higher complexity to lower complexity.

Now consider human consciousness in terms of complexity. The lowest level of complexity is to offer care and compassion to the 'self' only. Examples of increasing levels of complexity are to offer care and compassion to family, community, society, country, and eventually to all humanity; in other words, lower complexity equals lower inclusion and higher complexity equals higher inclusion, as the collective is a higher level of complexity than the individual. Based on this principle, it's reasonable to suggest that, if human beings are purely physical, then their natural tendency is to care for themselves rather than the collective.

Therefore, if life is a purely physical phenomenon, the present paradigm, which is dominated by SCAG, makes absolute sense. In other words, since selfishness is a natural force, it will always dominate in a human being if that human being is purely physical, and we have no choice but to accept SCAG as a natural phenomenon and get on with living in a world where the strongest survive and the winner takes all.

But what if life is not purely physical? What if there is a non-physical aspect to life?

Is there a non-physical aspect to life?

Six arguments suggest that life must be influenced by some force in addition to the physical – that is, in addition to matter.

Argument 1

For any system to exist, it must be influenced by two forces pulling in opposite directions in order to keep 'tension' on the system and, thus, keep it in existence. An analogy to explain this principle is the example of a clothesline, a system for drying clothes. The clothesline requires two forces pulling in opposite directions to keep it in tension; if one of these forces is removed, then the clothes line will collapse and the clothes will fall to the ground, reaching a state of increased entropy.

In a similar manner, life is a system that exists somewhere between minimum complexity (maximum entropy) and maximum complexity (maximum inclusion or maximum enlightenment). Therefore, if matter is the force pulling toward maximum entropy, then there must be another force pulling toward maximum complexity or maximum enlightenment. This force is likely not of a physical nature, so it is not yet understood by science.

Argument 2

Life can only evolve to either lower or higher complexity if the potential for that level of complexity already exists. Consider matter evolving to lower complexity, such as the example of a pipe returning to iron ore, at which point it reaches thermodynamic equilibrium with its surroundings. Now imagine the four forces of nature united (as in the Planck epoch), rather than existing as individual forces. In this scenario all matter, including the iron ore, would return to elementary particles (e.g., quarks, electrons) and finally to the source of elementary particles, which I have suggested is dark matter. However, matter could not possibly go to a level of complexity lower than elementary particles because the potential for that level of complexity does not exist.

In a similar manner, life could not evolve to a higher complexity unless the potential for that higher level of complexity already existed. In other words, a higher level of complexity of life cannot be 'created' by a purely physical process.

Argument 3

Humans have the capacity for reason, understanding, care, and compassion. Since none of these attributes are natural characteristics of matter, they must emanate from another source that influences human life.

Argument 4

Argument 4, an intuitive, rather than logical, argument, suggests that, if you look at life and nature in all its beauty, complexity, and diversity, it makes no sense to say that life is purely physical.

Argument 5

At a macro level, scientists accept that dark energy influences the behaviour of the universe because it causes the universe to expand at an accelerating rate. They also accept that dark matter influences the behaviour of galaxy clusters by keeping them gravitationally bound. However, at a micro level, scientists (geneticists, biologists) seem to deny that dark energy or dark matter have any influence on the origin and evolution of life. It is not logical to take the position that 96 per cent of the universe has no influence on life.

Argument 6

Evolution by natural means is unable to prevent the physical aspect of life from evolving to lower complexity, such as occurs with the ageing process, because matter has a natural propensity to return to its source, which is dark matter. Matter can achieve this goal only by returning along the path by which it emerged, finally reaching the state of quarks and eventually dark matter. In a similar manner, life cannot evolve to higher complexity by natural evolutionary means only: there must be a non-physical source at a maximum level of complexity to which life has a fundamental propensity to return.

Life: An Alternative View

These six arguments suggest the presence of a source or force other than matter and evolution by natural means that influences life and that, therefore, influences SCAG by balancing the force of matter.

Dark matter and dark energy, both of which are non-physical, are the phenomena from which the universe emerged, and all ordinary matter emerged from dark matter. In addition, ordinary matter has a propensity to return to lower complexity because of its primordial urge to return to its source. Hence, dark matter is the non-physical force that pulls life to lower levels of complexity, as occurs in the ageing process. Therefore, since there are only two non-physical forces known to science – dark matter and dark energy – it is logical that the force pulling life to higher levels of complexity must be dark energy.

If dark energy is responsible for the fact that the universe is still expanding at an accelerating rate, the natural tendency of dark energy must be to expand. Expansion implies greater inclusion and, therefore, greater complexity, so the natural propensity of dark energy is toward greater complexity.

If we accept the argument that dark energy influences life and that reason, care, and compassion are not characteristics of dark matter, then these traits must have emerged from dark energy.

Therefore, we can say that life is influenced by both dark matter and dark energy, the former pulling life toward lower complexity and the latter pulling life toward higher complexity. The physical and the non-physical are the two forces keeping life in existence: Death occurs when the physical aspect of life ceases to function, at which point the non-physical aspect is released to return to its source, which I believe is dark energy.

Three Unsupported Arguments

Three more arguments for this alternative view of life concern whether some other state emerged from dark energy, how dark energy and dark matter combine or unite to form life, and how mind is generated. These three arguments are not based on scientific concepts, but they may make it easier for us to visualise how dark matter and dark energy influence life, human minds, and human consciousness. Making these three points is a risky exercise, especially since science does not understand dark matter or dark energy or accept that dark energy has any influence on life. Erwin Schrödinger wrote the following circa 1944:

> "We feel clearly that we are only now beginning to acquire reliable material for welding together the sum total of all that is known into a whole; but, on the other hand, it has become next to impossible for a single mind fully to command more than a small specialized portion of it.
>
> I can see no other escape from this dilemma (lest our true aim be lost forever) than that some of us should venture to embark on a synthesis of facts and theories, albeit with second-hand and incomplete knowledge of some of them - and at the risk of making fools of ourselves"[2].

---

[2] Erwin Schrödinger, What is Life: with Mind and Matter and Autobiographical Sketches. New York USA: Cambridge University Press, 2012.

Nevertheless, if some of the ideas in this book will help humanity to move to a more enlightened paradigm, I feel it's worth the risk.

Physical entities like quarks, hadrons, leptons, nuclei of atoms, atoms, molecules, matter, galaxies, stars, planets, and moons all emerged from dark matter, so it is logical to assume that *dark energy entities* emerged from dark energy and formed a non-physical universe in the same way that the *physical entities* formed a physical universe. I will suggest that the emergence of *dark energy entities* was from levels of higher complexity (greater enlightenment) to levels of lower complexity. Therefore, the return of these entities is from lower complexity to higher complexity, which explains the natural propensity of *dark energy entities* to evolve to higher complexity.

My intuition is that the nature of these *dark energy entities* is that of 'individuated packets' of dark energy that are at a lower energy or vibration level than 'pure' dark energy and that there is a hierarchy of these *dark energy entities*, with the most complex at the top and the least complex at the bottom.

If my intuition is correct, every living organism is a union of matter and an individuated *dark energy entity*, and the *dark energy entity* is to the physical aspect of life what an electrical supply is to a computer: a human being is dead without a *dark energy entity* just as a computer is dead without an electrical supply. Since these *dark energy entities* are not physical in nature, they are not understood by science.

How do these *dark energy entities* unite with matter to form life? My first feeling is that this union is not superficial but is at the core of our being. However, science believes that DNA is at the core of our being, so it may be logical that the union of matter and *dark energy entities* takes place at this level.

Deoxyribonucleic acid (DNA) is a molecule that encodes the genetic instructions used in the development and functioning of all known living organisms. DNA molecules are normally double-stranded helixes that consist of two long biopolymers made of simpler units called nucleotides. The DNA is transcribed into mRNA, which carries coding information to the sites of protein synthesis called ribosomes. Here, the nucleic acid polymer is translated into a polymer of amino acids called a protein. mRNA is single-stranded, while DNA is double-stranded.

Could it be that one strand of DNA is encoded by dark matter for the physical being, while the other strand is encoded by dark energy for the *dark energy entity*? While nobody can either prove or disprove this suggestion, at a minimum, it is a powerful way of visualising how dark matter and dark energy are united in a living organism.

The next step is to suggest how this union of dark energy and dark matter might impact the human mind. At present scientists do not understand the nature of mind or how it is generated, but I propose a concept that at least makes it easy to visualise how mind is influenced by both the physical and non-physical aspects of our beings.

I suggest that mind is an energy that is generated in the core of the double helix of our DNA in the same way that a magnetic field is generated in an electrical coil. If one strand of our DNA encodes for the physical and the other strand encodes for the *dark energy entity*, mind is influenced by both the physical and non-physical aspects of our beings. Before mind is 'processed' by our brains, it is unconscious, and after processing it becomes conscious.

If we focus only on the physical aspect of our being, then it makes sense that selfishness will always dominate in our consciousness, and that, once we allow this to happen, greed will follow as surely as night follows day. Once selfishness and greed dominate in our consciousness, then corruption and abuse of power will almost inevitably occur when the opportunity presents itself.

If dark energy is the source of reason, intelligence, compassion, and love, and if we acknowledge and nurture this non-physical (dark energy) aspect of our beings, it will bring balance to our mind and move us from selfishness to love, care, and compassion. Therefore, focussing on and nurturing the non-physical aspect of our beings is the only 'antidote' for SCAG.

I have suggested some novel concepts that have the potential to change drastically our view of reality, but recall that science does not understand 96 per cent of that which constitutes the universe, since it does not understand the nature of dark energy and dark matter. Therefore, it is not logical for scientists to suggest that life is purely physical and that dark energy and dark matter have no influence on the origin and evolution of the universe or life in it. My view of reality incorporates 100 per cent of what constitutes the universe. The next chapter summarises the world view that results from this approach.

CHAPTER 4

Reality

Reality is the totality of all things, structures (actual and conceptual), events (past and present), and phenomena, whether observable or not. Reality is what a world view, whether it is based on individual or shared human experience, attempts to describe or map.

Figure 4.1 represents the view of reality that results from the ideas described thus far in this book. The figure shows that the universe is a dynamic phenomenon with a constant flow to and from each of its aspects. In other words, everything affects and is affected by everything else. This is the only view of reality that acknowledges the connectedness among all aspects of the universe, the continuous link between the source of the universe (a black hole) and human consciousness.

While a diagram such as that in Figure 4.1 is a good way to represent a complex process, it cannot provide the complete picture. For example, it gives

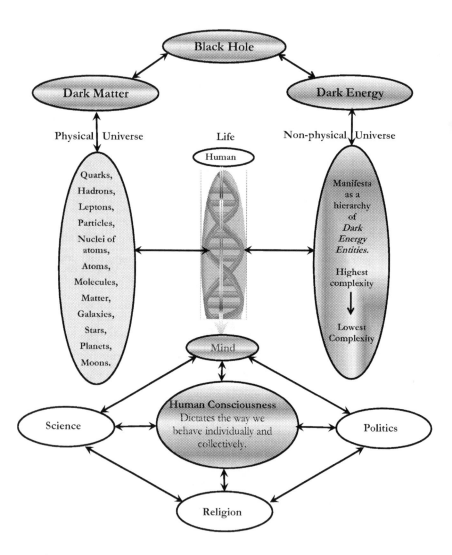

Figure 4.1: Reality from a Scientific Perspective

the impression that dark energy and dark matter exist separately, whereas they actually co-exist and permeate all of space. The physical universe, the non-physical universe and life also coexist.

Human Consciousness

What humans do and how they behave is mainly responsible for shaping the world in which we live. Why is our world in such a state of turmoil economically, socially, politically, and religiously? I believe that the answer to that question lies in the fact that we do not understand human consciousness, because it is human consciousness that drives what we do and how we behave as humans.

Many writers on the topic of human consciousness have described this phenomenon with varying degrees of success. However, that none of them have described it in a way that makes it possible for ordinary people to understand it or recognise its impact on how they live their everyday lives prompted Stuart Sunderland to write 'Nothing worth reading has been written about it.'

The main influences that inform human consciousness are mind, science, religion, and politics. There are other sources as well, such as art, music, literature, philosophy, and sport, but these are usually voluntary, while we cannot live in a society that is not influenced and/or controlled by politics and religion and informed by science. Of these four main forces, mind is the most influential, as it directly impacts human consciousness as well as religion, politics, and science.

It is clear, then, that mind plays a crucial role in shaping human consciousness, so it also plays a crucial role in shaping the world. Therefore, if our minds are influenced only by the physical aspect of our being, then a culture of SCAG will inevitably dominate our individual and collective consciousness. That is where humanity is 'stuck' at the present time and why there is so much poverty, starvation, conflict, and war.

CHAPTER 5

Religion

Many people depend on their particular belief in God to help them through the challenges of everyday life. However, it is important to differentiate between the faithful followers of religions and the leaders of their institutions. I respect the belief systems of the followers of all religions, but I believe that most of the leadership of religious institutions is dominated by the SCAG culture. They use the loyalty of their followers to propagate their own belief systems, which are not always for the benefit of humankind. In many instances, especially in relation to fundamentalists, the followers are brainwashed by the leaders, which can result in much division, conflict, and – all too frequently – bloodshed.

In general, religious institutions build up their own version of a 'religious god'. They then declare something along the lines of, 'If you belong to my religion and worship my god, you are saved; otherwise you are damned.' There are as many of these gods as there are religions, and these belief systems and their gods have been the root source of division and suffering for centuries. It is a futile exercise to criticise or question these belief systems, as that approach only adds fuel to the fire.

I believe that the term 'God' should be reserved solely for that which is the source of all. Therefore, there can only be one true concept of God because there can only be one source.

Some of the important facts and figures suggested by some of our most eminent scientists include:

- The universe is approximately 13.7 billion years old.
- The Earth condensed about 4.6 billion years ago.
- Life first emerged about 3.7 billion years ago.
- The human race (*Homo sapiens*) emerged about 195,000 years ago.

Since 'God' is a term coined by humans, it cannot be more than 195,000 years old. While the meaning of such terms can change over time and can vary across societies, the true *source of all* is singular and cannot change. Therefore, the term 'God' should be reserved solely for that which is the Source of All.

Although scientists can trace the evolution of the universe back approximately 13.7 billion years, they do not yet understand its source. However, it is logical to assume that, the farther back in time our understanding goes, the closer we will get to understanding the source and, consequently, to an understanding of God. Therefore, any concept we have of God must be compatible with our best understanding of that *source.*

Figure 4.1 identifies black holes, dark matter and dark energy as the Source of All, so if God is also the Source of All, then there must be a concept of God that is compatible with these phenomena. The only concept of God of which I am aware that is in any way compatible with the phenomena of black holes, dark energy, and dark matter is the concept of the Blessed Trinity: God the Father, God the Son, and God the Holy Spirit.

Black holes, dark energy, and dark matter are a scientific mystery, as scientists believe they exist but know nothing about their nature. The Blessed Trinity is a religious mystery made known to man by Divine Revelation. It is worth spending some time comparing these two mysteries.

God the Father and Black Holes

In Christian theology, God the Father is seen as the creator, law-giver, and protector of all that exists. He is viewed as immense, omnipotent, omniscient, and omnipresent, with infinite power and charity that go beyond human understanding. Some of the great mystics experience 'complete unity' when their awareness rests in God the Father. Ken Wilber explains non-duality (or unity) as 'The pure Emptiness of the Witness turns out to be one with every Form that is witnessed and that is one of the basic meanings of non-duality'.[3]

As discussed earlier (page 23), a black hole can also be viewed as 'complete unity' and, as scientists will acknowledge, it is a powerful phenomenon. In fact, it is not beyond the bounds of possibility that black holes and God the Father are actually the same phenomenon.

God the Son and Dark Matter

God the Son is the second part of the Trinity in Christian theology. Proceeding from the Father, the Son can be thought of as an immanent act of divine intellect or divine mind. (Immanence is the quality of any action that begins and ends within the agent, or put another way, it is initiated and consummated in the interior of the same being, so it may be considered a closed system.) Another way of looking at the Son is that He is the wisdom and power by which God is wise and powerful. To me, God the Son implies a physical reality and could be the source of the physical universe.

In the view of reality summarised in Figure 4.1, dark matter emerges from a black hole, which can be seen

---

[3] Ken Wilber,. (2000). *A Brief History of Everything.* Boston: Shambhala Publications, Inc., p. 206.

as analogous to the belief that God the Son proceeds from the Father. If dark matter may be the source of the physical universe, the concepts of God the Son and dark matter are synonymous.

God the Holy Spirit and Dark Energy

In Christianity, the Holy Spirit is the Spirit of God, the third part of the Trinity, which proceeds from the Father and the Son. The only way we can comprehend the Holy Spirit is to consider it an immanent act of divine will, just as we considered the Son an immanent act of divine intellect. Alternatively, just as the Son is the wisdom and power by which God is wise and powerful, so the Spirit is the holiness by which God is holy.

In the view of reality summarised in Figure 4.1, dark energy emerges from a black hole, which can be seen as analogous to the belief that God the Holy Spirit proceeds from the Father. Holiness also implies love and compassion, a characteristic I have attributed to dark energy. Therefore, in my view, God the Holy Spirit and dark energy are synonymous

I do not claim to understand the nature of God but am attempting to identify a single model that all humanity can embrace as an acceptable concept of God. If I can achieve that aim, it will help to convince the religious institutions and their followers that their 'religious gods' are not real and are the source of much division, suffering, and bloodshed.

However, even if humanity accepts the concept of God that I suggest, there will still be many questions that are beyond human understanding, including:

- Why do we exist?
- Why does anything exist?
- Why is there something rather than nothing?
- What was there before God?

In short, the religious mystery of God the Father, God the Son, and God the Holy Spirit will remain a mystery, as will the scientific mystery of black holes, dark energy, and dark matter.

Figure 5.1 presents the concepts in Figure 4.1 using religious terminology. The only difference between the two figures is the terminology:

- Black hole, dark matter, and dark energy have been replaced by God the Father, God the Son, and God the Holy Spirit.
- The non-physical universe has been replaced by the spirit world.
- *Dark energy entities* have been replaced by spiritual entities.

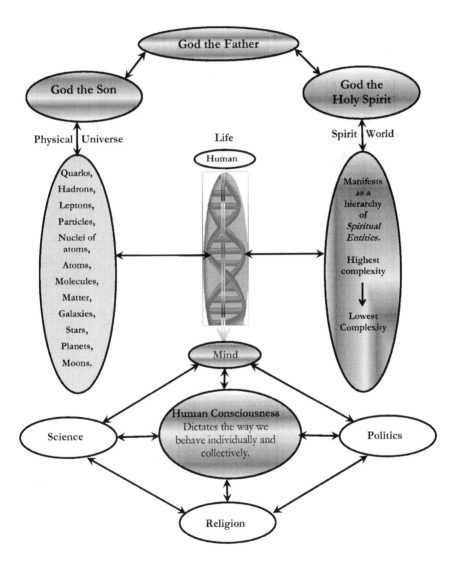

Figure 5.1: Reality from a Theological Perspective

Figure 4.1 and Figure 5.1 represent the same view of reality.

As was the case for Figure 4.1, Figure 5.1 cannot provide the complete picture. For example, Figure 5.1 gives the impression that God the Son and God the Holy Spirit exist separately, whereas they co-exist and permeate all of space. The physical universe and the spirit world also coexist. God the Son is the source of the physical universe, and God the Holy Spirit is the source of the spirit world.

Undoubtedly, equating God the Father to a black hole may cause problems for some people.

This is blasphemy!

The Irish playwright, George Bernard Shaw, proposed that 'all great truths begin as blasphemies'. In any case, I have not described the nature of God the Father, but have merely suggested a scientific concept that will give humanity a better understanding of God, the Source of All. I also see it as a way of convincing the general public and the various religions that we need to move forward from our entrenched positions to a more enlightened worldview.

A 'black hole' is not a very flattering way of describing God the Father!

True, but the only reason they are called black holes is that we cannot see them as they gather everything to themselves, and nothing can escape their gravitational pull; not even light. We can call them something else just as well.

There are billions of galaxies in the universe, so there must be billions of black holes. How can this be reconciled with the belief that there is only one God?

If you consider the water cycle analogy, there are seven seas and thousands of lakes of various sizes, but they are all connected via rivers and the ground water flow system, so there is actually only one body of water on the Earth. In a similar way, while there may be billions of black holes in the universe, they are all connected in their 'unity', so they are one body.

God the Father is a 'person' who will judge us on the last day. How can something like a black hole perform this function?

Is He a person as we understand the term, or is He an 'intelligence' or a 'consciousness' or 'pure love'? No one knows the answer to these questions. However, nobody knows what complete 'unity' is or how it behaves either. I assume that unity contains all matter, energy, light, intelligence, knowledge, wisdom, thoughts, ideas, love, and so on, as well as all their opposites (duality) that are manifested in the universe; and when all of this is converted or transformed into one homogeneous essence, one consciousness, one intelligence, we would then have 'unity'.

There are too many discrepancies between the religious view of God and the concept of black holes, dark energy, and dark matter.

If science can prove that complete unity exists within a black hole, then the number of discrepancies is not great. However, even assuming complete unity within a black hole, there are still two primary discrepancies. First, it can be inferred from some religious statements that God the Father is like a person: 'sitting at the right hand of God' or, 'On Judgement day, he will put all the good on his right hand side and all the wicked on his left and he will say to the wicked depart from me ye cursed into the everlasting fire prepared for you.' However, we don't know whether statements like these are metaphors used by Jesus and the prophets to get a point across or whether they should be taken literally. In my view, if God the Father is 'unity', then He cannot have a right and left. I also believe that, if there is complete 'unity' within a black hole, the same question arises for both the religious view and the scientific view regarding God the Father: Is God the Father a person or an entity? If the answer is that He is a person, then the religious view would be that He is a person who exists in heaven, and the scientific view would be that he is a person who exists in a black hole. On the other hand, if the answer is that He is a spiritual entity, then the religious view would be that He is a spiritual entity whose essence is heaven, and the scientific view would be that He is a spiritual entity whose essence is a black hole. Maybe heaven and a black hole are synonymous.

The second discrepancy relates to the Holy Spirit. The religious view of the Holy Spirit is that He proceeds from the Father and the Son, whereas the scientific view suggests that dark energy emanates only from black holes. The religious view on this point is surrounded in mystery, theology, and dogma, so it is difficult to reconcile the two viewpoints.

In summary, there are many aspects of the religious mystery of God the Father, God the Son, and God the Holy Spirit and many aspects of the scientific mystery of black holes, dark energy, and dark matter that we may never understand. However, these mysteries should not prevent us from trying. We cannot advance our understanding of God from the scientific viewpoint until science develops a better understanding of black holes and that which emanates from them.

CHAPTER 6

The Evolution of Human Consciousness

Introduction

Since the time of Descartes, philosophers have struggled to comprehend the nature of consciousness. Stuart Sutherland wrote: 'Consciousness is a fascinating but elusive phenomenon: it is impossible to specify what it is, what it does, or why it has evolved. Nothing worth reading has been written on it.'[4] Other definitions of consciousness include 'a sense of one's personal or collective identity, including the attitudes, beliefs, and sensitivities held by or considered characteristic of an individual

---

[4]   Stuart Sutherland. (1995) The International Dictionary of Psychology, 2nd Ed. New York: Crossroad,

or group';[5] 'sentience; awareness; the ability to experience or to feel; wakefulness; having a sense of selfhood; or as the executive control system of the mind';[6] and 'anything that we are aware of at a given moment forms part of our consciousness, making conscious experience at once the most familiar and most mysterious aspect of our lives'.[7]

I consider consciousness to be the sum of our experiences, both real and imaginary, as understood by our minds, which are influenced by both the physical and non-physical aspect of our beings. In this regard, what we experience can be influenced by the physical aspect of our being, which pulls us toward SCAG (lower complexity), or by the non-physical aspect of our being which pulls us toward love and compassion (higher complexity). Therefore, mind influences and is influenced by what we experience, which influences our consciousness in turn.

In the remainder of this chapter I will use religious rather than scientific terminology because religious

---

[5] The American Heritage Dictionary of the English Language, 5th edition, 2013 by Houghton Mifflin Harcourt Publishing Company, Boston.

[6] G. Farthing. (1992). *The Psychology of Consciousness*. Englewood Cliffs, *NJ*: Prentice Hall.

[7] S. Velmans & S. Schneider. (2007) *The Blackwell Companion to Consciousness*. Malden, MA: Blackwell.

terminology is more common; for example, 'spiritual entities' (or spirits) is more commonly used than is 'dark energy entities.' I consider the following terms to be interchangeable, as they express the same aspects of reality:

- Black hole/God the Father
- Dark matter/God the Son
- Dark energy/God the Holy Spirit
- *Dark energy entities*/spiritual entities/individuated spirits/spirits
- Universe of *dark energy entities*/spirit world.

In the interest of simplicity, I use either masculine or feminine pronouns, rather than joint pronouns like 'he/she'.

General Description

When a child is born, his mind is essentially a blank, apart from survival instincts and the rich evolutionary history up to humans, all of which come as part of the package. Immediately after birth, the child starts to learn and become aware, first of his physical boundaries, then his emotional boundaries. He later becomes aware of his responsibilities to his caregivers, then his family, later still his social group, then his country, and after that to all humanity. These are all expanding states of consciousness. They bring with them increasing responsibilities and duties of care, which require the child/adult to embrace each new stage with an increasing level of love, compassion, and responsibility. While the journey from birth to enlightenment is a continuous, gradual process, it can be broken down into ten stages. Each stage includes all the previous stages, in much the same way that matter includes molecules that include atoms that include particles that include quarks.

How Do We Negotiate a Stage?

There are three steps involved in negotiating any stage: awareness, attention, and responsibility.

Awareness

If you are not aware of a stage, you can do nothing about it. Awareness normally comes naturally but can also be prompted by your group or society. The awareness for a stage usually starts developing towards the end of the preceding stage.

Attention

Once you become aware of a stage, you must give it your attention, analyse it, and become fully informed about it. Children and adults will, by nature, give the required attention to each stage, but 'nurture' sometimes gets in the way, as I discuss later.

Responsibility

You must then identify with this new level of awareness, enfold it within your being, and take responsibility for the increased level of love and compassion that the new level of awareness demands.

A good analogy for this three-step process is learning to ride a bicycle. When a child is about four years old, she becomes aware that she would like to ride a bike. This awareness may have been prompted by a friend having learned to ride a bicycle, her parents having bought her a bike for her birthday, and so on. Prior to that time, riding a bike wasn't in the child's awareness; she was probably too busy learning how to crawl, walk, and talk. However, once riding a bike came into her awareness, the child gives it attention, and after much practice, which includes falling and being frustrated, the child eventually acquires the ability to ride a bicycle. This newfound ability can be dangerous unless the child learns that she must be responsible when riding a bike, which means wearing a helmet, not going too fast, obeying the rules of the road, and so on. Once she has learned to ride a bike as described, that ability is then enfolded into her being, and barring a serious accident or traumatic experience, she will never lose that ability, as once you have successfully negotiated a particular stage, you can never go back; you will always have that stage of awareness enfolded within your being.

I have divided the evolutionary journey into ten stages. Each stage must be successfully negotiated before we can negotiate the next one, so we cannot skip a stage. Although we can be aware of and start negotiating a higher stage before we have successfully negotiated the stage below, we cannot complete the higher stage until we have successfully completed the lower stage, as each higher stage includes all the lower stages.

The Four Quadrants of Consciousness

The individual's evolutionary process has two aspects that move to higher complexity in tandem: the physical and the non-physical. This process can be compared to that of a computer, which has both hardware and software dimensions. In the 1980s, computer hardware was simple, speeds were slow, and storage capacities were small; accordingly, the software applications were basic. Today we have PCs that have processing speeds in excess of 4 GHz, memories (RAM) of over 4 GB, and storage capacities in excess of 500 GB. Therefore, we can have software applications that are much more powerful and complex. Just as the software evolved in tandem with the hardware, the physical (hardware) and the non-physical (software) of the individual evolve in tandem.

But the individual's physical and non-physical aspects do not evolve in isolation. In tandem with the evolution of the individual is the evolution of the collective, which also has physical and non-physical dimensions: the social and the cultural, respectively. Figure 6.1 summarises these four quadrants in relation to humans.

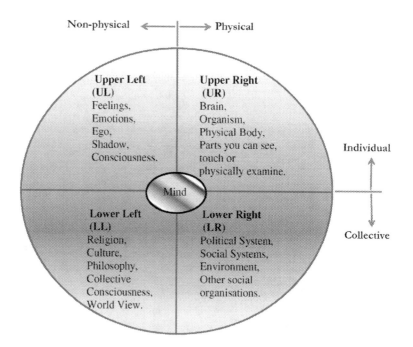

Figure 6.1: The Four Quadrants of Consciousness

Figure 6.1 shows 'mind' in the middle, as it is with mind that we see, feel and understand the four quadrants of the individual's physical and non-physical, the cultural and the social.

These four quadrants interact with and influence everything we do and think. For example, each quadrant influences how I play a game of golf.

Influences of UL

- I must first have the thought that I have the option to play a game of golf. If I had been born before the Middle Ages, I would not even have the thought, as golf didn't exist at the time.
- I must have the desire to play.
- How I feel inside, as every golfer knows, will have a major impact of the quality of my game.
- My emotions will influence my game.

Influences of UR

- My physical body – my physical strength, fitness, flexibility, and so on – has a major influence on my golf game.
- My brain also has an influence: eye-to-ball co-ordination, choice of club for various shots, judgement of slopes on greens, and so on.

Influences of LL

- My desire to play golf is greatly influenced by the culture and the belief systems of the society in which I grew up. For example, if my society considered golf only a game for women, I probably would not wish to play the game.

- My knowledge of golf and the best way to play it is greatly influenced by the people I talk to and play with and by how the books and newspapers I read describe the game.

Influences of LR

- My society provides the golf course on which I play. If there were no courses, I couldn't play the game.
- My society also provides the golf clubs, bag, trolley, balls, and other equipment; without them I could not play.
- My country's political system influences the availability of good golf courses through, for example, policies regarding funding and planning and how golf may be used to promote tourism.

The Evolution of Consciousness in the Human from Birth Onward

The stages of an individual human's evolution starts at birth and can be broken down into ten stages:

H1 – Physical
H2 – Emotional

H3 – Ego
H4 – Reason
H5 – Reflection
H6 – Integrated Personality
H7 – Nature
H8 – Spirit World
H9 – Causal
H10 – Non-dual

The ten steps from H1 to H10 make up the 'human journey, or the H journey. Why do I call it a journey? When we are born, we are generally unaware of our surroundings, our family, our society, our country, and everything else. The journey of life is to bring all these aspects of the human experience into our awareness with love and compassion for all.

Each stage in the H journey brings with it a different view of the world, the self, and others – in other words, a different *worldview*. In the H journey, the order is irreversible, and stages cannot be bypassed, so the individual must go from stage H1 to stage H2, to stage H3, and so on. Each stage includes all the previous stages.

Being at any particular stage is not 'wrong'; for example, there is nothing wrong with being at stage H4, as this stage is just as important as any higher stage. Even if you are at a higher stage, you must have the lower stages enfolded within that higher stage, and even if a society as a whole is at H6, there will always be individuals at H1 to H5, as there will always be babies born who have to start their development at H1. One must respect where every other being is in relation to his or her development. For example, that you are at H5 and your neighbour is at H4 does not mean that you are better than he is or inferior to somebody who is at a higher stage than you are.

When you are born, you are at H1. Then you quickly evolve up to the stage to which your society has evolved, at which point you generally 'join the herd'. The rate of your development from then on is usually dictated by your society. The requirement is that you keep moving and do not stagnate at any particular stage.

Before I describe each stage, I should explain a few important concepts:

- It is with our mind that we see and understand our individual and collective reality.
- During stages H1 to H4, the child's consciousness moves to shallower and narrower perspectives (Figure 6.2). In addition, the child does not initially complete any of these four stages fully; for example, when the child moves to H3, he does not enfold and take responsibility for his emotions. Thus, if a child at H3 gets angry, he blames his sister/father/mother, or if he is feeling bored, he claims 'This place is really boring,'

and so on. H1 to H4 is a process of learning about oneself. It can be compared to a trainee car mechanic who, in order to understand how the engine works, first takes it apart and analyses each part separately. However, when he has taken apart and analysed the carburettor, he does not responsibly put it back on the engine but leaves it on the floor while he disassembles and analyses the next part. Another way of looking at this concept is that, during this familiarisation process, the child moves to shallower and shallower waters until eventually he can comfortably stand up with his head above water, take a few breaths, and get ready for the enfolding process that occurs at H5.

- During H5 our mind is still dominated by the physical aspect of our being as we continue to learn, understand, and navigate our way through the physical world. Our spiritual aspect is also evolving during this time and influencing our behaviour to the extent that it is being nurtured and respected. On completion of H5 we are a fully integrated person and aware that we should extend love, care and compassion to our fellow man. **It is at this point that humanity is stuck.**

- We cannot negotiate H6 until we become aware of and focus on the spiritual aspect of our being.

- The word 'environment' refers to everything outside the particular aspect that is being addressed or the aspects that have been addressed and integrated up to that time.

- The age bands for the various stages shown in Figure 6.2 are approximate, as every child and every generation are different.

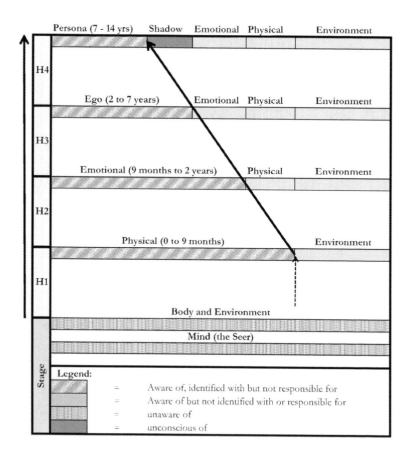

Figure 6.2: The Evolution of Human Consciousness, H1 to H4

Each stage is described under three headings:

Goals: The goals that will be achieved as the stage is negotiated.

Developments: The aspects of the individual that will undergo major development as the stage is negotiated.

Worldview: The view of the world that usually accompanies the successful negotiation of the stage.

H1 – Physical (0 to 9 months)

Goals: I call this first stage *Physical* because it is during this period (birth to 9 months) that the child develops his physical boundaries.

Developments: When a child is born, she cannot differentiate among mind, body, and environment; in other words, these aspects of life are all fused together, and she cannot tell the difference. Immediately after birth, the baby undergoes rapid development, although the rate of this development is significantly influenced by the level of love, care, nurture, and stimulation she receives from her caregivers during this time. At birth, the baby's mind is not a complete blank; instead, it includes all the rich evolutionary history up to humans, all the survival instincts that have evolved, and perhaps also any 'baggage' or Karma that might come with that particular spirit. However, at this stage the baby has no awareness of herself or her environment.

While the self at this stage is more than just physical, it is still predominantly oriented to the lowest and most basic dimension of all, the gross or physical world. The physical self and the physical world are still fused, or to put it another way, they are not yet differentiated. The infant can't tell the difference between his own physical body and his surroundings. Some people suggest that the baby is 'one with all' at this stage and, therefore already at H10, but the baby does not transcend subject and object (non-duality); it simply can't tell the difference between them. The development during this stage is selfish and narcissistic because the physical world and the physical self are still fused. During H1 the child starts differentiating his own physical body from that of its surroundings by a series of experiments and experiences. For example, when the baby bites his mother's finger, it doesn't hurt, but it does hurt when he bites his own finger, so the child realises there is a difference. This differentiation is usually completed between 5 and 9 months, at which point the baby has established realistic boundaries of his physical self.

Worldview: The worldview at this stage is referred to as archaic, a term that represents the totality of the rich evolutionary history up to humans, which lives on in each of us as part of our complex individuality.

H2 – Emotional (9 months to 2 years)

Goals: After about nine months the infant has realistic boundaries of her physical self but has not yet established the boundaries of her emotional self, which is still fused or identified with those around her, particularly her mother. The goal for this stage is to establish realistic emotional boundaries.

Development: All aspects of the child continue to evolve and develop, significantly influenced by the level of love, care, nurture, and stimulation the baby receives from her caregivers during this time. The development at this stage is still egocentric and narcissistic because the child has no real emotional boundaries, so she cannot differentiate between her emotional self and the environment. Somewhere around 12 to 24 months, the emotional self differentiates from the environment, at which point the child realises that she is a separate individual who is not really fused with her mother. This realisation brings with it its own fear and heralds the start of the 'terrible twos'. At this stage the infant is beginning to wake up to the fact that she is a separate and sensitive emotional being, but since she cannot yet take on the role of 'other' (or empathise) and does not yet realise that everybody else is also an emotional being, she still thinks selfishly.

Worldview: The worldview at this stage is called magic. The child lives in a world that she can manipulate by magic.

H3 – Ego (2 to 7 years)

Goals: The goal for the child at this stage is to continue to acquire a better understanding of himself by continuing to differentiate to narrower and shallower aspects of himself. During H3 the child differentiates to and identifies with the ego (Figure 6.2).

Development: At this stage the child begins to transcend the emotional level by identifying with the representational mind, which consists of images, symbols, and concepts. Images emerge at about 7 months, while symbols dominate from about age 2 to 4 years, and concepts from about age 4 to 7 years.

- An *image* looks more or less like the object it represents. For example, if I show a child a picture of a car, he will relate that picture to 'Daddy's car' since it looks very similar

- A *symbol* represents an object but may not look like the object at all, so understanding a symbol is a much more difficult cognitive task. If I write the word 'car' on a piece of paper, it is a much more difficult task for the child to relate this word to 'Daddy's car'.

- *Concepts* represent an entire class of objects. For example, the word 'car' represents all cars and not just 'Daddy's car'.

During this stage the child becomes aware of impulses and emotions, and begins to recognise symbols and concepts. He then starts to talk and write, which opens up a whole new world. ***Worldview:*** The worldview at this stage is called mythic. The child realises that magic doesn't work anymore, and he hopes that mythical forces, such as fairy godmothers, tooth fairies, Santa Claus, and other special forces will help him achieve his aims. This mythic worldview of stage H3 continues into stage H4.

H4 – Reason (7 to 14 years)

Goals: The goal at this stage is to continue the differentiation of the self to a point at which the child is comfortable putting the parts back together again, thus forming a whole and integrated person. By the end of H4 (approximately 14 years of age), the child will have differentiated to and identified with his persona and may have removed from his conscious mind that which he found too traumatic to remember; this is sometimes known as the shadow self.

Developments: By the end of stage H3, the child is able to reason. During stage H4 he develops the capacity to form mental rules and to take mental roles, so the child learns to take the *role of other* and to see that his view is not the only view in the world.

Worldview: The child's worldview switches from the egocentric stance of the previous stages to sociocentric, extending love and compassion to 'my friends, my group, my tribe, my religion,' but no farther.

H5 – Reflection (14 years to ?)

Goals: During stages H1 to H4, the child differentiates to shallower levels, but he doesn't really complete step 3 (enfold and take responsibility) for any of the stages. For example, if a child at H3 gets angry, he blames his sister/father/mother, and if he is bored, he claims, 'This place is really boring,' and so on. In the analogy of the trainee mechanic, he has taken the engine apart in stages H1 to H4, and he must now put it back together again in H5. The goal for H5 is to integrate and take responsibility for the physical, emotional, ego, and shadow aspects of the self. From the perspective of the four quadrants (Figure 6.1), the teenager must integrate the two upper quadrants, the individual's physical and non-physical. By the end of H5, the young adult usually feels 'happy in his own skin'. Figure 6.3 depicts stage H5 graphically.

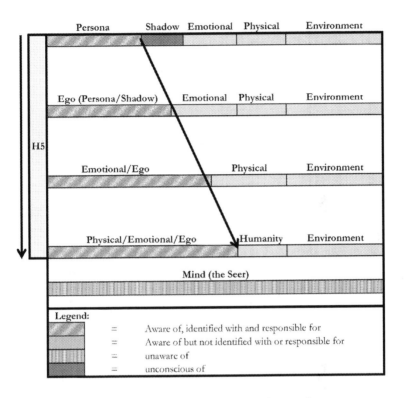

**Figure 6.3: The Evolution of
Human Consciousness, H5**

Development: The development during this stage is, to say the least, hectic. During H4 the child develops the capacity to form mental rules and to take mental roles, but at H5 she develops the potential to reflect and form her own views on existing rules and norms. At this stage, the teenager has the ability to analyse and think about the physical world but also to analyse and criticise the norms, rules, and ideals of her society. The teenager can now imagine different possible worlds and grasp 'what ifs' for the first time. To paraphrase George Bernard Shaw: During stage H4 children see things as they are and ask why. During stage H5 teenagers dream things that never were and ask why not.

At this time teenagers are full of imaginings, idealistic thoughts, and dreams. They will challenge the status quo and all authority, particularly that of their parents, whose belief systems they see as conservative and outdated at best and utterly stupid at worst. It is the age of revolution, where teenagers think they can change the world. It is also the age of sexual blossoming, when hormones run wild. Any parent who has raised teenagers will readily admit that the 'terrible twos' is a walk in the park compared to this stage. Because teenagers can now think about thinking, they can start to judge the roles and the rules that they previously accepted without question. They can agree or disagree with the norms of their society and, therefore, transcend them to a degree. By the end of H5, the teenager can usually see the humour in this old joke: When I was 15 I thought my parents knew nothing; by the time I was 25, I was impressed by how much my parents had learned in ten years.

Worldview: If you are lucky enough to negotiate this stage successfully, you will have gone from socio-centric at H4 to a world-centric and global perspective. While you have not yet reached the stage at which you can extend love, care, and compassion to the global community, you still want to know what is right and fair – not just for you and your people, but for all people.

However, completing H5, where you believe that care and compassion should be extended to all humanity, is difficult. **This is where humanity is stuck.**

Comment

Stage H5 usually starts to emerge at around the age of 14, and under ideal conditions should be completed by the time the individual is in her mid-twenties. However, life is not ideal, and few people complete this stage successfully. At the extreme, when a child has been the victim of serious traumatic experiences during its formative years, such as abandonment, physical or sexual abuse, bullying, to name but a few, it will take much longer to negotiate H5. These horrendous experiences leaves the victim with shame and guilt as well as understandable anger, resentment, and bitterness toward the perpetrator, some or all of which the teenager may have 'forgotten' and banished to her shadow self. All these feelings will incapacitate the victim, leaving her unable to love and respect herself, and without love and respect for yourself, it is not possible to love and respect anyone else.

People who have had such serious traumatic experiences during childhood usually require professional help to bring such experiences back into their awareness so they can forgive the perpetrator and let go of the memories. As counter-intuitive as it may seem, genuine forgiveness is the only way to let go of these memories. An unfortunate aspect of this type of experience is that it is often passed on from generation to generation (at a collective level), and a vicious cycle is then set up. The only way for this vicious cycle to be broken is when a critical mass of people negotiates H5 and moves to H6 so love, care, and compassion can be extended to all people and not just the privileged few.

It is difficult for any modern society to enable a critical mass of people to negotiate H5, as the religious, political, and public service institutions of nearly all modern societies are afflicted by SCAG:

- In most modern capitalist democracies, the SCAG culture is blossoming at all levels of political and public service institutions. If we analyse the financial near-collapse in 2008 honestly, we have to admit that the SCAG culture is largely to blame.
- In emerging countries dictators accumulate vast fortunes for themselves and their families at the expense of their people, whom they control by force and terror.
- Many religious institutions are controlled by leaders who are interested only in maintaining the status quo, particularly their own positions of power and control over their followers.

- In many societies there is a massive increase in religious fundamentalism, where misguided religious leaders brainwash their young followers to carry out horrendous acts of violence in the name of their god.

These are but a few examples that demonstrate that all modern societies are to some degree afflicted by this SCAG culture. Unfortunately then, when our young men and women reach their early twenties, they quickly realise that their teenage idealism gets them nowhere and that, if they are to be successful in their careers, they must 'join the herd', and the herd in all modern societies is controlled by SCAG.

The reality that the battle for the hearts and minds of our young people is being won by SCAG has serious consequences for our society at both an individual and a collective level. For individuals to live by the SCAG philosophy, they must suppress the 'cry of the soul', which they do by means of drug abuse, alcohol abuse, immoral behaviour, and other decadent and self-indulgent practices that provide instant physical gratification but that in the longer term destroy them physically, emotionally, and spiritually. At a collective level we are left with a society that extends care and compassion to nobody, so the vulnerable in our society are ignored.

Thus, with the continued evolution of the information and knowledge base, people who are not sufficiently evolved to use this power for the benefit of humanity gain power and use it for their own selfish agendas. The effect on our society is analogous to giving terrorists the knowledge to develop nuclear weapons.

This short commentary briefly explains why no modern society has so far developed a critical mass of people who have negotiated H5 and moved to H6, where love, care, and compassion are offered to all people.

H6 – Integrated Personality

Goals: The goal for this stage is to extend love and compassion to all humanity. To achieve this goal it is necessary to develop a fully integrated personality by integrating the four quadrants (the individual's physical and non-physical, the cultural, and the social) and the mind (see Figure 6.1). The upper two quadrants (the individual's physical and non-physical) were integrated during H5, and in H6 the goal is to extend this integration to include the two lower quadrants (the social and cultural) plus the mind. At this stage the observing self begins to transcend the mind and body; thus, the mind and body become integrated to create for the first time a fully integrated personality. When this happens, the observing self becomes centred in the spirit, which then becomes the 'seer' or the 'witness within' (Figure

6.4). In other words, the mind becomes dominated by the spiritual aspect of the being which empowers us to extend love care and compassion to all humanity.

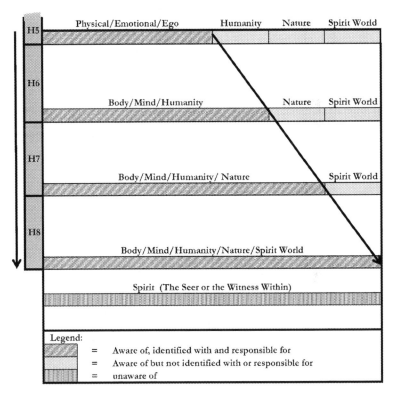

Figure 6.4[8]: The Evolution of Human Consciousness, H6 to H8

---

[8] The top line in Figure 6.4, which represents the completion of H5, shows Physical/Emotional/Ego. These are all aspects of the integrated body/mind, and therefore in the lower lines, the words 'body/mind' are used to represent all of these aspects.

The physical, emotional, and ego are all aspects of the integrated body/mind. While up to this stage the mind is dominated by the physical aspect of our being, at this point we begin to go transpersonal as the mind becomes dominated by the spirit. H6 is a transition stage, and it is only on the successful completion of this stage that we can say that our observing self is centred in our spirit. By the end of H6 we have gone transpersonal, so stages H7 to H10 are known as the transpersonal stages.

Development: The main development at this stage is learning to embrace the spirit and allowing it to dominate and inform the mind because it is virtually impossible to extend love and compassion to all humanity from the perspective of a mind that is dominated by the physical aspect.

Loving all humanity is extremely difficult:

- You may have worked hard all your life to support your spouse and family, but your neighbour, who has never worked a day in his life, lives off the welfare system and seems to have a better standard of living than you do.
- Every day you come across people in all walks of life who are corrupt and seem to buy their way to success, while you must earn every penny the hard way.

- Every day you read of religious fundamentalists who carry out the most horrendous acts of violence on society and whose beliefs, from your perspective, are evil.

These examples show how difficult it is to extend love and compassion to all humanity when your mind is dominated by the physical aspect of your being. The only way to extend love and compassion to all humanity is to allow your mind to be dominated and informed by the spiritual aspect of your being. Doing this is not like flicking a switch but requires a long and difficult process of soul-searching and doubt, and making this transition requires honesty and authenticity. Therefore, H6 can be an unhappy place because while you have not yet fully gone transpersonal: you have experienced all that life can offer and find it to be unfulfilling and profoundly meaningless.

In summary, extending love and compassion to all humanity is counter-intuitive when your mind is dominated by the physical aspect of your being, because the physical aspect naturally draws you toward selfishness. However, as the evolution of consciousness continues and more depth is disclosed, the observing self starts to transcend mind and body by shedding the lesser identifications with both in order simply to witness them, which is why both mind and body are experienced as an integrated self. At this point, the observing self begins to transcend mind and body and to go transpersonal.

The only way to eliminate the present injustices and contradictions is for the collective society to extend love, care, and compassion to all people, which can happen only when a critical mass of people negotiate H6. In an ideal world, H6 should begin to emerge in each individual's mid-twenties, but for two reasons this does not happen. The first reason is that the religious, political, and public service institutions of nearly all modern societies are afflicted by SCAG which prevents a critical mass of people completing H5. The second reason is that, at around this time, nature demands that we do our bit to ensure the survival of the species. Therefore, from our late-twenties to our mid-forties, much of our attention and energy is directed toward getting a job, falling in love, building a home, and rearing a family. It is usually not until all this is accomplished that we have time to reflect on the greater meaning of life or, as some people might say, we reach the 'mid-life crisis'. However, when a critical mass of people in any society has negotiated H6, it will become easier for this stage to emerge and it will be negotiated at progressively earlier ages by succeeding generations.

The reluctance of science to acknowledge the non-physical (spiritual) aspect involved in H6 and subsequent stages has serious implications for the evolution of human consciousness to a more enlightened worldview where care and compassion will be extended to all people. The reluctance of scientists to acknowledge any phenomenon that cannot be verified by observation or experiment is understandable, but the fact that scientists do not understand at least 96 per cent of reality has not prevented them from positing that dark matter and dark energy exist and influence the universe at a macro level.

In a similar manner, it should be equally obvious that dark energy influences life and is the force that is responsible for the evolution of human consciousness to higher levels. If science acknowledged this reality, human consciousness would quickly evolve to a more enlightened worldview.

Worldview: The world view at this stage is world-centric and includes all humanity. At H5 we learned to love ourselves, and we became aware that we should love all humankind. However, at that stage we only 'talk the talk', whereas at H6 we learn to 'walk the walk'.

All subsequent worldviews are world-centric, but they come with greater and wider levels of inclusion (Figure 6.4).

H7 – Nature

Goal: The goal for H7 is to extend love and compassion to all sentient beings and to all of nature.

Development: The main development at this stage is an expansion of consciousness to include all sentient beings and nature. Up to this point, the journey has taken us from matter to body, to mind and to spirit, and in each case the observing self transcends a lesser and shallower dimension and identifies with a deeper and wider consciousness. In stage H7 we complete our identity with the complete physical universe, or God the Son. In H8 and H9 we open up fully to the Holy Spirit, which is the source of the soul, the observing self, the witness within.

Traditional science and many religions do not acknowledge the transpersonal stages, so this is where the mystical, contemplative, and yogic traditions pick up the story. They push deeper and deeper, not just into their own awareness but into the source of awareness itself. These traditions are to the spirit world what the scientific community is to the physical world. Few people, including writers on the topic, have negotiated the transpersonal stages, particularly H9 and H10, so we must rely on the great mystics to inform us about them. The great mystics say that, when they quiet their minds, allowing their awareness to go beyond the ego and the individual self and rest in the source of awareness itself, they become aware of a subtle essence that pervades all reality and that this reality is the fertile ground of all that is. This reality is pure spirit.

Stage H7 is the first of the transpersonal stages, when we move from the ordinary physical reality into the transpersonal domains. The defining characteristic of this level is an awareness that is no longer confined exclusively to the individual ego or mind. Your sense of self can dissolve, and you can identify with an entire physical experience, such as watching a beautiful sunset, where you become part of the experience. Suddenly there is no one looking: you and the sunset are one; you have become part of the sunset. In other words, there is no separation between the subject and the object, between you and the entire natural world. At H7, nature is a part of you, so you treat it with respect.

To somebody who has not reached H7, all this might sound weird. However, it is possible for people at earlier stages to have peak experiences of nature mysticism: drug-induced experiences or those due to extreme stress, near-death, or moments of sexual passion. However, such peak experiences are only temporary, as it is necessary to evolve sequentially through all the previous stages to hold this level permanently.

Worldview: The worldview at this stage is world-centric, but it's a world-centric view that has moved from including only all human beings (H6) to including all sentient beings and nature (H7).

H8 – Spirit World

Goal: The goal at this stage is to expand the awareness to include the spirit world. On completion of H7, you extended love and compassion to the complete manifest physical world, but the challenge in H8 is to extend that love and compassion to include the spirit world. The spirit world is comprised of a hierarchy of spiritual entities that emerged from pure spirit, as shown in Figure 4.1 and Figure 5.1:

• Your deity, whatever you perceive that deity to be: Christ, Mohammad, the Buddha, or others
• Angels, archangels, guardian angels, spirit guides, and so on
• Saints and other highly evolved souls who have been through the 'Earth experience' many times
• Less evolved souls who may have to pay another visit to Earth to heal some negative experience or action
• Young souls
• Lost souls
• Spirits of sentient beings, such as animals, birds, and fish
• Spirits of living objects of nature, such as trees, plants, and flowers

Development: At this stage you transcend nature mysticism and identify with deity mysticism through a union with your deity by whatever name. A Christian might see this identification as one with Christ, an angel, or a saint; a Buddhist might see it as identification with the Buddha; a Jungian might see it as an archetypal experience of the self; and so on. This is where confusion regarding God usually arises, but God should be recognised as the Source of All, while deities are highly evolved spiritual entities whose spirits may be one with the Holy Spirit and, therefore, one with God. However, being God and being 'one with God' are two separate concepts.

One's individual background, cultural background, and social institutions all influence how one interprets this experience. This is the beginning of stage H8. As you negotiate this stage, you develop the ability to be aware of various spiritual entities and forces, which may bring with it various degrees of fear and even terror.

Of course, when you move to any new and higher stage, there is always fear of the unknown, at least initially. For example, when a child leaves the security of the home and goes to school for the first time, she usually feels anxious and frightened. When we move to H8, we are for the first time aware of the awesome power and myriad forces of the spirit world, which may be terrifying at first.

Worldview: The worldview at this stage is world-centric. It includes not only the physical manifest world but also the spiritual world.

H9 – Causal

Goal: The goal at this stage is to become aware of the Holy Spirit, the Great Spirit, dark energy, or whatever name you prefer to call this infinite, unmanifest essence.

Development: At stage H8, your awareness expanded to include the complete manifest universe, which includes both the physical world and the spirit world. The challenge at stage H9 is to move your centre of observation from the manifest world to the unmanifest world in order to experience God the Holy Spirit. This supreme spirit is the source of infinite bliss, love, and compassion, in much the same way that our spirit is the source of our own love and compassion. This state is often described as an infinity experienced in a fullness of 'being'. It is a unique, pure-energy state of being that cannot be contained by any manifestation; in fact, it is prior to any manifestation. This experience can best be explained by a passage from the Chandogya Upanishad, translated by Swami Krishnananda: 'There is a subtle essence that pervades all reality. It is the reality of all that is and the foundation of all that is. That essence is all. That essence is the real. And thou, thou art that.'

At this stage, the witness within (the seer) actually becomes one with the Holy Spirit ('thou art that'), who is the creative intelligence of all manifest dimensions. When your awareness rests in the Holy Spirit, you can still see the real world, including your ego, your mind, and your body, but you are not any of these things. Instead you are pure awareness itself, and you exist as that awareness.

In previous chapters I suggested that the Source of All (God the Father, God the Son, and God the Holy Spirit or black holes, dark energy, and dark matter) existed prior to the Big Bang. The great mystics confirm this insight by saying that the pure seer is prior to life and death, prior to time, prior to space, prior to manifestation, prior to the Big Bang itself. This state is called 'causal' because it is the support, cause, or creative ground of all junior dimensions.

Worldview: The world view at this stage is still world-centric, but it includes not only the complete manifest world (physical and spiritual) but also God the Holy Spirit.

H10 – Non-dual

Goal: In H1 to H7, we brought God the Son into our awareness with love and compassion. In H8 and H9, we brought God the Holy Spirit into our awareness. When we move to H10, we can see both God the Son and God the Holy Spirit from a place of unity or non-duality. In other words, the witness within completely vanishes and becomes everything that is witnessed. The contemplative traditions believe that at stage H10 we experience total unity, a union with God the Father. Some mystics experience this as pure emptiness, a key concept in the ontology of Mahayana Buddhism. The phrase 'form is emptiness; emptiness is form' is perhaps the most celebrated paradox associated with Buddhist philosophy. Ken Wilber explains unity or non-duality as follows: 'The pure Emptiness of the Witness turns out to be one with every Form that is witnessed, and that is one of the basic meanings of non-duality.'[9]

Development: At H10 we integrate the witness within with the manifest and unmanifest worlds. We are now resting in and experiencing unity, which cannot be a discrete state but is the reality of all states since it includes and enfolds all other states. All states continue to arise in the moment, but there is nobody watching; there is just the brilliant display of great perfection. You have moved from the causal to the non-dual, which is the very nature of God before He expresses Himself as God the Son and God the Holy Spirit. It is not some state you bring about through effort but the condition of all experience before you do anything to it. Only extremely evolved beings can experience this non-dual state.

Worldview: 'Worldview' is meaningless at this stage because in unity there is no other; there is nothing to describe. There is only unity, and that unity is all – God is All.

---

9   Ken Wilber. (2000). *A Brief History of Everything.* Boston, MA: Shambhala Puiblications Inc., p. 206

Some Comments on the Human Journey

Most people can relate to some of the stages of the human journey. Any parent who has raised children will easily understand H1 to H5, and all people will remember some of stage H3 and onward in their own lives. The stage of collective awareness that is reached by even the most enlightened cultures at present is the mid to latter stage H5, where care and concern are extended from the individual to his country ('from me to my country') but no further. A few cultures are struggling to complete H5, where a world-centric and global perspective starts to emerge, or perhaps I should say that a few pioneering souls are leading their people through stage H5, moving from socio-centric to world-centric. They want to know what is right and fair, not just for them and their people, but for all people.

Navigating the transpersonal stages (H7 to H10) requires meditation and contemplation since we must look within ourselves. Unfortunately, most people at stage H5 ridicule such practices. However, even though you may not understand these stages, it is important to keep an open mind and to appreciate that, even though we may not understand the contemplative traditions, it is they that will eventually lead us to enlightenment. The contemplative traditions are to the spiritual world what scientists are to the physical world.

Lines of Development

There are several lines of development or evolution that humans must negotiate, including the cognitive, moral, interpersonal, spiritual, emotional, and aesthetic. While the descriptions of the H Journey have assumed, for the sake of simplification, a balanced or average development across all of these lines of development, Figure 6.5 demonstrates how a person can be at a different level on each line.

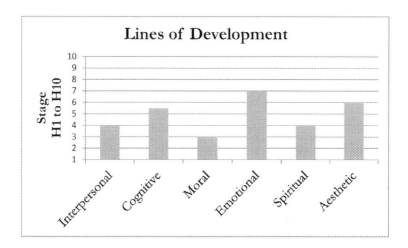

Figure 6.5: Human Consciousness, Lines of Development

Chapter 7

Summary

It is clear that the human race is a highly dysfunctional family. If any other family were behaving in such a manner, the children would be taken into care and the parents/guardians would be charged with neglect and cruelty. The human race seems incapable of running our little world such that all of its members are treated with compassion and respect.

Our world is afflicted with extreme poverty, starvation, conflict, war, diseases that are curable, and so on. All of these problems are avoidable, yet they have plagued humanity for millennia. Why can't we move to a more enlightened view of reality?

The fundamental problem is selfishness, an innate quality in every human being. Rather than looking at selfishness as an evil or 'sinful' quality, we must accept that it is a natural characteristic of the physical aspect of our being and consider two options for dealing with this naturally occurring characteristic:

- Accept that life is purely physical and that there is no non-physical aspect to it. Therefore, accept that selfishness will always dominate in our being since the natural propensity of matter is to evolve to lower complexity. When selfishness dominates in our consciousness, greed naturally follows, and then when the opportunity presents itself, corruption and abuse of power raise their heads. Since, in this view of reality, there is no 'antidote' for selfishness, we have no choice but to accept our lot and get on with living in a world where SCAG will always dominate, where the strongest survive, and the weak are ignored and neglected.

- Accept that there is a non-physical aspect to life that, when it is acknowledged and nurtured, has the potential to counteract the physical characteristic of selfishness and empower humanity to move to a more enlightened worldview, where love, care, and compassion are extended to all, not just the privileged few.

Therefore, the big question is this: how can humanity move to a more enlightened paradigm?

What must be done?

Humanity must accept that there are two aspects to life: the physical and the non-physical. The non-physical can be described as dark energy or spiritual. When humanity accepts this view of reality, it can nurture and embrace its dark energy/spiritual aspect, which will influence minds and counteract the propensity of the physical aspect to evolve toward selfishness. It will then be possible for humanity to continue on its evolutionary journey by completing stage H5 and moving through stage H6, where care and compassion are extended to all humans.

How can this be done?

Both religion and science have a role in ensuring the acceptance of both the physical and non-physical aspects of our beings.

Religion

Many religious institutions have for centuries said that there is a spiritual aspect to life that is the source of our love and compassion. Why has this message not helped humanity move to a more enlightened paradigm? Why should I expect that essentially the same concept of life presented in this book should be any more successful?

When Christial[10] was asked for the greatest commandment, He replied, 'Love the Lord your God with all your heart and with all your soul and with all your mind. This is the first and greatest commandment. And the second is like it: Love your neighbour as yourself. All the Law and the Prophets hang on these two commandments' (Matthew 22:36–40).

Therefore one of the two central messages in the teachings of Jesus Christ is to 'love your neighbour as yourself'. However, this is precisely where humanity is stuck – two thousand years after Christ. Why have our Christian churches failed so miserably to convince humanity to embrace the message of Christ and move to a higher level of enlightenment over the past two thousand years? SCAG in the hierarchy of all the main Christian Churches has stripped these institutions of the necessary moral authority to enable them to convince the general public of the importance of Christ's message.

The central problem is that, as any religious movement grows, it becomes 'corporate' and under the control of individuals who are mainly interested in power and control. SCAG then grows to a point where

---

[10]   The references made to Christ and Christianity are made for purposes of brevity only and could equally be applied to the other prophets and their respective religious institutions.

it dominates in the institution, and at that point the survival of the institution becomes more important than the fundamental teaching of the prophet on which it is based. For example, the Roman Church seems to have been more interested in its own survival than in living by the fundamental teaching of Jesus. The excesses of the Roman church have been evident in many historical examples:

- The Spanish inquisition and the wider Christian inquisition in the late fifteenth century
- The sale of indulgences for monetary gain, which led to the Protestant Reformation in the sixteenth century (While Martin Luther disputed the claim that freedom from God's punishment for sin could be purchased, as the Protestant movement grew, it also became 'corporate' and eventually came under the influence of SCAG.)
- The failure of the Catholic Church to address adequately the problem of child sexual abuse by its members (The hierarchy seems to have been more interested in preserving its power than it was in protecting children from this horrendous crime.)
- The obsession of the Catholic church with its man-made rules and regulations regarding sex outside marriage, masturbation, homosexual practices, attending mass on a weekly basis, and

so on, while ignoring the fundamental message
of Christ to love your neighbour

- The many reports of major corruption at the
centre of the Catholic Church over the last fifteen
hundred years

As people became educated and more aware of the
prevalence of SCAG in the Christian churches, they
lost faith in their teaching. This is the main reason why
the message of Jesus has not grown sufficiently in the
consciousness of the general public over the past two
thousand years to counteract the influence of SCAG.

Science

On the other hand, science seems to have been largely
unaffected by SCAG, primarily because all the great
scientific discoveries have resulted from the personal
endeavours of dedicated scientists who loved their
work – people like Galileo, Kepler, Newton, Maxwell,
Curie, Einstein, Bohr, and Feynman, to name but a few.
Because scientific research did not become 'corporate',
it was less likely to be influenced by SCAG. Having
said that, science is now becoming more 'corporate'
and more exposed to the culture of SCAG, mainly
because of the huge profits available in pharmaceuticals,
genetically modified crops and food stuffs, information
technology systems and software, and so on. For the

time being, however, many parts of science are still largely unaffected by SCAG.

Fortunately, in the mid-sixteenth century, when the Christian churches were suffering the worst influences of SCAG, the scientific revolution started when Nicolaus Copernicus published *De revolutionibus orbium coelestium* (*On the Revolutions of the Heavenly Spheres*). Since then, scientists have taken our understanding of the universe to new levels.

At a macro level, scientists now contend that dark energy and dark matter permeate the universe, comprising approximately 96 per cent of the 'whole', and that dark energy is responsible for the expansion of the universe, which is still expanding at an accelerating rate. I believe that, with continued research into dark energy and dark matter, scientists will conclude that dark energy also influences life, that dark energy is responsible for the expansion of human consciousness, and that it is the source of our love and compassion.

When this time comes, science can declare that life is influenced by both the physical and spiritual aspects of our being. This reality can then be taught to our children as a scientific fact, rather than as a religious belief (since teaching religious beliefs in schools can be divisive). This view of reality will then seep into our political, public service, religious, and financial institutions. In a relatively short time, a critical mass of people will embrace this new view of reality, and we will be on our way to a world in which all people will be afforded love, care, and compassion. Over time, this will put an end to wars, poverty, and starvation and will herald a new beginning for humanity.

This is Humanity's path to freedom from selfishness, corruption, abuse of power, greed, poverty, starvation, conflict, and war.

Bibliography

Braden, Gregg. (1997). *Awakening to Zero Point*. Bellevue: Radio Bookstore Press.

Bradshaw, John. (1990). Home Coming. New York: Bantam Books.

Brennan, Barbara Ann. (1993). *Light Emerging*. New York: Bantam.

Carroll, Sean B. (2006). The Making of the Fittest. New York: W. W. Norton & Company.

Chopra, Deepak. (1993). Ageless Body, Timeless Mind. New York: Three Rivers Press..

de Chardin, Teilhard. (1975). *The Phenomenon of Man*. New York: Harper Colophon.

de Mello, Anthony. (1992). *Awareness*. New York: Doubleday.

DiSalle, Robert, 'Space and Time: Inertial Frames'. *The Stanford Encyclopedia of Philosophy* http://plato.stanford. edu/entries/spacetime-iframes/

Dyer, Wayne W. (2004). *The Power of Intention*. Carlsbad, Ca:Hay House.

Grof, Stanislav. (1993). *The Holotropic Mind*. San Francisco: Harper Collins.

Grof, Stanislav. (1998). *The Cosmic Game*. Albany: State University of New York Press.

Hanh, Thick Nhat. (1991). *The Miracle of Mindfulness*. London: Rider.

Hawking, Stephen and Mlodinow, Leonard. (2005). *A Briefer History of Time*. London: Bantam Press.

Hawking, Stephen and Mlodinow, Leonard. (2010). *The Grand Design*. London: Bantam Press.

Jung, C. G. (1995). *Memories, Dreams, Reflections*. London: Fontana Press.

Jung, Carl G. (1990). *Man and His Symbols*. Middlesex, England: Penguin/Arkana.

Maltz, Maxwell. (1960). *Psycho-Cybernetics*. Englewood Cliffs, NJ: Prentice Hall.

Murchie, Guy. (1999). *The Seven Mysteries of Life*. Boston: Mariner Book, Houghton Mifflin Company.

Myss, Caroline. (1997). *Anatomy of Spirit*. New York: Three Rivers Press.

Newton, Michael. (2001). *Journey of Souls*. St Paul, Minnesota: Llewellyn Publications.

Numbers, Ronald L. (2006). *The Creationists*. Cambridge, Ma: First Harvard University Press.

O'Leary-Hawthorne, John. (2008). *Substance & Individuation in Leibniz*. New York: Cambridge University Press.

Peck, M. Scott. (1990). *The Road Less Travelled*. London: Arrow.

Purcell, Brendan. (2011). *From Big Bang to Big Mystery*. Dublin: Veritas.

Redfield, James & Adrienne, Carol. (1997). *The Celestine Prophecy*. New York: Warner Books.

Thompson, Bert. (1999). *The Scientific Case for Creation*. Montgomery, Alabama: Apoletics Press Inc.

Velmans, M. & Schneider, S. (2007). *The Blackwell Companion to Consciousness*. Malden, MA: Blackwell.

Walsch, Neale Donald. (1996). *Conversations with God*. New York: G.P Putnam's Sons.

Wilber, Ken. (1993). *The Spectrum of Consciousness*. Wheaton, Illinois: The Theosophical Publishing House.

Wilber, Ken. (1996). *A Brief History of Everything*. Dublin: Gill & Macmillan Ltd.

Wilber, Ken. (1996). *The Atman Project*. Wheaton, Illinois: The Theosophical Publishing House.

Wilber, Ken. (2000). *Sex, Ecology, Spirituality*. Boston: Shambhala Press.

Wilber, Ken. (2001). *A Theory of Everything*. Boston: Shambhala Publications Inc.

Wilber, Ken. (2007). *Integral Spirituality*. Boston: Integral Books.

Zukav, Gary. (1990). *The Seat of the Soul*. London: Rider.

Index

U

Understanding 35
Union 39
Unity 24, 49, 50, 55, 56, 93
Universe 12, 35, 43, 48
Unmanifest 28
Unmanifest world 92

W

War 6, 46
Water cycle 21, 55
Weak Nuclear. *See* Four forces
Weak nuclear force. *See* Four
    forces
Wilber, Ken 50, 94
Wisdom 50
Witness within 84
Worldcentric 78, 87, 89, 91
Worldview 71

Y

Yogic traditions 88